THE POCKET ESTIMATOR

Serves as a practical guide and reference for field personnel in developing material quantities and labor hours for estimating the cost of all disciplines of building construction.

Published in Conjunction with

THE BUILDING ESTIMATOR'S REFERENCE BOOK

22nd EDITION

FRANK R. WALKER COMPANY
Publishers   **Chicago**

TABLE OF CONTENTS

PREFACE

The Pocket Estimator was developed as a handy reference manual for field supervisors to assist them in their daily decisions relating to material purchasing, crew assignment and production, cost control, and the general accepted methods of construction.

Virtually all of the information has been extracted from The Building Estimator's Reference Book and, in instances where more detailed information is required, the user is directed to the appropriate section of the Reference Book. In instances where labor rates are provided, these are the trade prevailing rates established by the Department of Labor for the Metropolitan Area of the District of Columbia, therefore, they should be adjusted to the local prevailing trade wage rates. Manhours or production rates provided are for average conditions and should be adjusted to compensate for unusual conditions should they exist.

The Pocket Estimator is considered the finest, most practical, and accurate source of information available to the field supervisor and, when properly used, should contribute to the profitability of any project.

William H. Spradlin, Jr.
Senior Editor

Wheaton, Maryland
1985

CHAPTER 1

TOOLS AND EQUIPMENT

The owning, rental or use of tools and equipment subjects the contractor to compliance with specific safety requirements which may be imposed by cities, states and the Federal Government. It is not unusual to detect considerable conflict or variance in the degree of protection required under the several sources of regulations. In such instances, it has been held that the contractor is required to comply with the maximum requirement.

A national safety guide is the Occupational Safety and Health Act (OSHA) established by the Department of Labor in 1970 and all tools and equipment utilized on a construction project should meet these standards. Purchase Orders or Rental Agreements should contain a specific clause relative to this compliance.

Failure to meet the provisions of the safety laws can subject the contractor to the penalty of fines and costly litigation.

DETERMINING HOURLY RATE

Whether the equipment is rented or owned, the estimator should charge for it at a fixed hourly rate. This rate would include delivery, set-up and dismantling charges plus the basic rental. This would then be divided by the working hours that the equipment stayed at the site. In some cases one might elect to charge only for the hours that the equipment was in use at the job site. The remainder of the time would then be charged to office overhead. Operating costs would be added as a separate item only for the actual in-use hours.

Equipment rental rates vary greatly throughout the country, depending upon local practices and conditions. They may also vary depending upon the use to which the equipment is placed; such as hard service in rocky soil or for use in sand.

It is general practice in the industry to base rates upon one shift of 8 hours per day, 40 hours per week, or 176 hours per month of a 30 consecutive day period.

The lessor usually bears the cost of repairs due to normal wear and tear on non-tractor equipment. On rubber-tired hauling and tractor equipment, the lessee may be required to bear all costs for maintenance, and repair.

The equipment rented should be delivered to the lessee in good operating condition, and is expected to be returned to the lessor in the same condition as delivered, less normal wear.

National average rates quoted herein are f.o.b. the lessor's warehouse or shipping point. The lessee pays all freight or delivery charges from shipping point to destination and return. Lessee also pays all charges for unloading, assembling, dismantling and loading where required.

Rentals are usually paid in advance and subject to the terms and conditions of the lessor's rental contract. Rates do not include contractor's insurance, license fees, sales taxes or use taxes.

MATERIAL HANDLING

Hand shovelling of bulk materials will vary as to whether the load is transferred at the same or lower elevation or raised to a higher one. The following table is based on the amount one man can handle in one hour under dry conditions.

	Even Transfer		Upward Transfer	
Sand	3	tons/hr	2.6	tons/hr
Stone	2.12	" "	1.8	" "
Gravel	2.3	" "	2	" "
Top soil	2.6	" "	2.35	" "

Sizes, Weights and Capacities of Contractor's Wheelbarrows

Struck Capacity Cu. Ft.	Heaped Capacity Cu. Ft.	Approx. Weight Pounds	Type of Materials Used for
2½	3½	74	Dry Materials
3	4	78	Dry Materials
3½	4¼	75	Concrete-Mortar
4	5	79	Concrete-Mortar
4½	6	83	Concrete-Mortar

It requires about 150 shovels (No. 2 shovel) for one cu. yd.

Chutes and Conveyors.—The cheapest energy the estimator can plan for is gravity, and he should see the job is laid out so that material distribution will be by gravity in so far as possible. Straight line chutes can be rented in ten foot sections for $10.00 to $12.00 per week or may be built up out of plywood at the site. Prefabricated metal vertical chutes will cost $20.00 to $25.00 per foot erected. Gravity steel roller conveyors can be purchased in 10' lengths for around $175.00 for 16" widths. Powered conveyor belts will run two and a half times as much plus one must add for operating costs.

Portable Belt Conveyors.— Some applications in the construction field require the use of portable belt conveyors of various sizes and capacities. Their flexibility does not limit the user to a fixed receiving or discharging point, and when used in series, this flexibility is still further increased to carrying of materials over long horizontal distances, over obstructions, around corners, etc.

Portable conveyors are available in boom lengths to 120 ft. and belt widths of 18", 24", 30", and 36". Swivel wheels permit building of high capacity radial stock piles or discharge into any one of several bins or hoppers. Both single and multi-deck vibrating screens may be used at the discharge end to separate the material into two or three sizes. Feeder-trap and feeder-hopper units mounted at the tail end, permit charging by front end loaders, dozers, cranes, draglines, shovels or other conveyors.

HOISTING TOWERS

Tubular steel hoisting towers are extensively used by contractors in all parts of the country for hoisting all types of building materials and for concreting, because of the ease with which they are erected and dismantled, their great strength and stability, as well as their economy.

Tubular hoisting towers are designed for handling material and concrete, with or without chuting. They are built in any height and with one or more compartments. Their many advantages both in construction and in performance, their durability and low maintenance, and their remarkable adaptability to various jobs, make them ideal as low cost equipment for contractors.

A tubular hoisting tower consists of a bottom section, the necessary number of intermediate tubular sections, each 6'-6" in height, and a top section containing the cathead with sheaves.

All intermediate sections of a given type of tower are identical and the top and bottom sections can be attached to any one of them.

The single and double (two compartment) towers are classified as follows:

	Heavy Type	Light Type
Size of legs	3-in.	2-in.
Maximum height	350'-0"*	201'-0"
Live load capacity	5,000 lbs.	3,200 lbs.
Maximum size concrete bucket	35 cu. ft.	....

*Higher towers require special material.

The size of the cage platform in the single tower is 5'-10½" x 7'-6½". For double towers the size of one cage platform will be 5'-10½" x 7'-6½" and the other platform will be 5'-10" x 7'-2".

Erecting and Dismantling Steel Tubular Towers.—When estimating the cost of erecting and dismantling a steel tubular tower, the number of man-hours involved depends upon the height of the tower and the general conditions involved. For instance, it will take considerably longer to erect a tower 208'-0" high than one 65'-0" high, not only in time consumed but also in man-hours per section, taking into consideration the amount of material involved as well as the height to which the equipment must be raised.

Time given includes erection, dismantling and handling in man-hours per lin. ft. of height.

Light Type Single Tubular Towers .... 1-1.5 hr.
Light Type Double Tubular Towers .... 1.5-2 hr.
Heavy Type Single Tubular Towers .... 1.5-2 hr.
Heavy Type Double Tubular Towers .... 2-2.5 hr.

Prices of light or heavy duty hoisting towers should be obtained from the manufacturer of such equipment, at the time that they are needed. Quotations for rental charges on a monthly basis can also be obtained for a specific project. A 50' tower will rent for about $1,500 per month, with each additional 10' about $40 extra per month.

SCAFFOLDING

Tubular Steel Scaffolding.—The most versatile type scaffolding is galvanized steel tubing with a male fitting on one end and a female fitting on the other to form interlocking units. These units are then coupled together with a patented coupling device. It can be adapted to a variety of uses and its simplicity has found general acceptance among the building trades. Basically, it consists of four parts; interlocking steel tubes of various sizes, a base plate or caster, and an adjustable or right angle coupler. All parts are galvanized as a permanent protection against rust.

This galvanized carbon steel tubing or pipe is approximately of 2" o.d. (1½" Standard Pipe) and weighs 2¾ lbs. per ft. Other sizes are 2½" o.d. (2" Standard Pipe Size) which weighs about 3¾ lbs. per ft.

Standard lengths for tube are 6'-0", 8'-0", 10'-0", 13'-0", 16'-0" and 20'-0", and for pipe are 6'-0", 8'-0", 10'-0" and 13'-0". Other sizes may be obtained but those mentioned are the sizes generally used by the building trades. Each section is fitted with a bayonet locking device.

This locking device between members is permanently affixed to the tubes, which method insures a perfectly rigid attachment which has been tested to a load of 20,000 lbs. in tension. The assembled tube and fittings make up a complete unit so designed as to have flush joints.

Standard couplers are made of drop-forged parts, except for the steel hinge pins on which the catch bolts hinge. The most widely used couplers are 2" x 2", both adjustable and standard, and 2" x 2½", both adjustable and standard, and 2½" x 2½", adjustable and standard. This type of scaffold is the most versatile of all built-up scaffolds.

Costs of purchasing or renting exterior tubular steel scaffolding are quoted on the square footage of wall area actually covered, and may be obtained from manufacturers of the equipment. Low buildings will cost about $40.00 per hundred sq. ft. installed (rental). Costs for buildings over 5 stories will cost about $60.00 per hundred sq. ft. of surface installed (rental).

Costs for purchasing or renting interior tubular steel scaffolding are quoted on the number of cubic feet actually occu-

TOOLS AND EQUIPMENT

pied by the scaffolding and may be obtained from the manufacturer or supplier of the equipment.

Sectional Steel Scaffolding.—Sectional steel scaffolding or prefabricated scaffolding is readily assembled and installed and is well adapted to various kinds of building and maintenance work. Various sizes of 5'-0" wide frame units, trusses and ladder scaffold components, make it simple to erect a suitable and a rigid, safe scaffold under any conditions.

This is a lighter type scaffold than the tube and coupler type and has an approved working height limit of approximately 125'-0".

The spacing between frames may be either 3'-0", 5'-0", 6'-0", 7'-0" or 10'-0", and the sizes of the frames may vary from 3'-0" to 10'-0" in height, thereby allowing for a wide variety of uses. Braces are attached to the frames by means of wing nuts or automatic locks.

Suspended Scaffolding

The most economical method of scaffolding for masonry work on buildings with a structural steel or reinforced concrete skeleton framework is suspended scaffolding. The working platform is supported by regularly spaced scaffolding machines, each consisting of a pair of cable drum hoist mechanisms with connecting steel putlog, suspended by steel wire rope from steel or aluminum outrigger beams anchored to roof or floor framing above.

Standard width machines accommodate a 5'-0" width of solid planking. Machines are available to accommodate 8'-0" wide decking to permit wheeling of material on scaffold platform. Machines are also available for mechanized material handling with the inboard cable drums set back from the wall to provide a 5'-0" wide platform for material handling traffic and stockpiles on one side and a separate 20" wide working platform for masons on the other. All machines are available with or without steel frames to support overhead protective planking. For maximum safety, overhead planking should be used, together with toe boards and guard rails along outer edge of platform.

Scaffolding machines are usually spaced 7'-0" apart, permitting the use of 16'-0" scaffold plank. A ratchet device on the drum hoist mechanisms provides positive locking while still permitting easy raising or lowering of scaffold to maintain bench- high location of work for maximum efficiency and best workmanship.

Outrigger beams are usually anchored to roof construction 7'-0" on centers, if building is not over 150'-0" high, the maximum lift capacity of scaffolding machines. For structures exceeding 150'-0" high, one or more additional installations will be required on lower floors. There are various methods of anchoring beams to building framework, such as securing with steel straps to anchor bolts set in concrete slabs or beams; bolting to inserts set in concrete: and anchoring to beams or joists with U-bolts and bars through holes in concrete slabs.

SIDEWALK BRIDGES

Where construction takes place adjacent to heavy traffic areas and it is impracticable to re-route sidewalk traffic into the street, a sidewalk bridge must be erected to protect the public. Rigid standards must be met in the design and erection of these structures.

Construction is as follows: 4½" o.d. steel columns set on heavy timber mudsills support 8" steel I-beams bridging the sidewalk. Beams are securely attached to columns by means of special steel column head castings. Columns are tied together longitudinally with two rows of pipe girts with an occasional diagonal pipe brace as required. Wood timbers, cut to a taper and set to slope towards the street, are fastened to the cross beams with U-bolts to support 6" steel beams, spaced approximately 16" on center, running the length of the sidewalk bridge. Two layers of 2" or 3" planking are then laid over the beams, the bottom layer running laterally and the top layer longitudinally. A parapet for the street side may be added if desired.

Most contractors lease sidewalk bridges on an erected and dismantled basis and the supplier usually quotes a lump sum figure for a specific job. Before submitting a bid, a firm quotation covering this work should be obtained, where required.

For preliminary estimating purposes only, and average conditions, sidewalk bridges may be figured as follows:

Erection and dismantling, per lin. ft.	$20.00-$36.00
Sidewalk bridge rental, per lin. ft., per month...	4.75
Erect and dismantle parapet, per lin. ft.	9.00
Parapet rental, per lin. ft., per month.................	2.00

AIR COMPRESSORS

Two and more tools can be operated from the same compressor. The rating of the compressor should be the delivery, but leaking hoses, poor tools and adverse job conditions could lower the cfm output. In some gang operations not all units will be in operation at the same time so that a unit sized for two may be adequate for three, providing the compressor is set up for more than 2 hose connections.

The following list will give some idea of air requirements for the more commonly used tools:

Breakers		Drills	
35 #	– 30 cfm	Air	
60 #	– 40	1"	– 35 cfm
80 #	– 50	2"	– 50
		4"	– 60
Diggers		Wagon	
20 #	– 20 cfm	3"	– 160 cfm
25 #	– 30	3½"	– 200
35 #	– 35	4"	– 250

Hammers
Jack
    25 # – 40 cfm
    45 # – 90
    75 # – 160
Chipping
    light – 20 cfm
    heavy– 30 cfm

Rivet
    ⅝ ” – 25 cfm
    ⅞ ” – 35

Saws
Circ.
    12” – 50 cfm
Chain
    30” – 90

    36” – 140
    48” – 160

Tampers
    35 # – 30 cfm
    60 # – 40
    80 # – 50

Vibrators
    2½” – 25 cfm
    3” – 45
    5” – 80

Wrenches (Impact)
    ⅝ ” – 15
    ¾ ” – 30
    1½ ” – 75

Compressor rentals will vary as to the condition and rated size of the equipment, but should be available for following approximations:

fuel	cfm	per day	per week	per month
gas	85	$ 34	$110	$ 260
”	100	40	110	270
”	125	40	110	270
”	150	60	190	415
”	175	75	200	425
diesel	250	100	245	570
”	375	130	300	700
”	600	150	400	1,100
”	750	225	700	1,800
”	1000	280	925	2,800

Tools, hoses and other attachments must be added separately.

CHAPTER 2

SITE WORK

CSI DIVISION 2

02010 SUBSURFACE EXPLORATION

When preparing estimates on any structure involving foundations or earthwork, it is often necessary to make borings or test holes to ascertain the nature of soil which will be encountered at various depths below the ground surface, and to determine whether or not water will be found and at what depth. Frequently this information is shown on the working drawings to aid the foundation contractor in preparing his bid.

There are many methods used to investigate soil conditions for foundation purposes. Some are good while others produce inadequate or dangerously deceptive results.

Hand Auger Boring Method.—The hand auger boring method, satisfactory for highway explorations at shallow depths, can be used only for preliminary investigations for foundations. Auger borings are used in cohesive soils and cohesionless soils above ground water. Depths up to 20 feet may be reached. Soil samples obtained through this method are disturbed to such an extent that little or no information is furnished about the character of the soil in its natural state. Where the auger hole is filled with water, sample disturbance is particularly critical. Auger holes are most useful for ground water determination.

Hand auger borings generally cost from $4.00 to $6.00 per lin. ft. depending on the soil, the amount of boring necessary, and the diameter of the hole which will vary from 4" to 8".

Split Spoon Sampling.—A split barrel sampler known as a split spoon is frequently used for obtaining representative samples. This method is described in detail in ASTM Method D-1586.

In securing split spoon samples, the drill hole is opened and cleaned to the desired sampling level by drilling bits and wash water, or alternatively the hole may be drilled with a regular power auger. Drill casing or drilling mud is used when it is necessary to prevent the soil above the sampling level from closing the drill hole. A drive pipe is attached to the upper end of the drill rod and a slip jar weight placed on the drive pipe.

The sampling spoon is then driven into the soil by repeated blows of a drop hammer. The number of blows required to drive the spoon are recorded on the logs. The spoon is then brought to the surface and the sample removed, classified and placed in a glass jar.

The ASTM method specifies a 2-inch OD x 1-⅜ inch ID split spoon driven 18" to 24" by a 140 lb. weight falling 30 inches. Various other size split spoons, drop weights and length of drops are employed by some companies. The split spoon method is most useful in granular or dense soils.

Thin-Wall Tube Sampling.—The method of sampling which has been found to give the least disturbed samples is the thin-wall tube method. These samples are most suitable for testing. This method is described in detail in ASTM Method D-1587.

In this method the hole is opened and cleaned as in the split spoon sampling method, but a thin-wall tube, usually with an outside diameter of 2", 3" or 5" is then attached to the end of the drill rod and lowered to the bottom of the boring. The tube is then pushed into the soil by a hydraulic piston arrangement with a rapid and continuous motion.

The thin-wall tube is then raised to the surface, cleaned, labeled, and sealed to prevent loss of moisture. It is then taken to the laboratory for testing. Here it is usually cut into short lengths and the sample in each length ejected and tested.

Samples are normally taken at 5' intervals or every time a soil change is expected as determined by watching the overflowing wash water and feeling the resistance to penetration of the drilling bit, irrespective of the distance between changes.

Cost of drilling and sampling is usually between $6.00 and $16.00 per lin. ft. depending on the depth, diameter, and difficulty of drilling, plus on and off charges for the equipment and men. Laboratory testing is performed at an additional cost, depending on what tests are needed—for density, moisture content, compression or mechanical analysis.

Core Borings.—In addition to soil samples, information is frequently wanted on the nature or thickness of underlying rock. Core borings are made for this purpose with rotating coring tools such as diamond drills, shot drills, or carbide bits. Continuous samples are generally recovered for examination and further testing. ASTM Standard D-2113, Diamond Core Drilling for Site Investigation, should be called for in performing core borings.

The cost of making core borings will depend upon the location of the work, quantity, and the difficulty encountered in getting to the holes.

The drilling costs per lin. ft. should be estimated at from $18.00 to $22.00 per foot of drilling with the lower costs on jobs of larger size. To the above prices must be added the on and off charges for the equipment and men, and testing charges.

02100 CLEARING

Stripping and Storing Top Soil.—Following site clearing, the first item usually considered is stripping top soil. Top soil is generally specified to be stripped from all areas of the site to be occupied by structures, paving, walks, etc., and to be stockpiled for later use in finish grading. To compute the quantity of top soil to be removed, take the dimensions of the areas and add an amount necessary to clear the top line of slope for general excavation to each dimension. The product of these dimensions multiplied by the depth of the top soil will give the amount of top soil to be stripped and stockpiled.

Cutting Down and Removing Trees

The cost of cutting down and removing trees will vary with their size and the method used in cutting and disposing of them. If their roots run through the portion to be excavated it will increase the labor cost of excavating and removing the soil.

Chain saws are obtainable operated by gasoline or electric power or compressed air, and with 24", 36", 48" and larger saws, and rent for around $30.00 per day.

Using a chain saw with 3 men in the crew cutting trees, removing branches and cutting the trees into 2'-0" lengths, should average as follows:

Approx. Diam. of Tree	Approx. Height of Tree	Labor Hours	Extra Labor Removing Stump
8" to 12"	20'-0" to 25'-0"	2¼ to 3	8
14" to 18"	30'-0" to 40'-0"	3 to 4	10
20" to 24"	45'-0" to 50'-0"	5 to 6	12

A faster and more economical method of removing trees is by using a bulldozer or loader for this work.

A WORD OF CAUTION—Always watch for falling limbs. In the case of pines, if the whipping action is too great, the top may snap off and kick back on to the machine or operator. Also as the tree begins to fall, the machine must be backed up quickly so that the upturning root system will not come up under the machine with the possibility of damaging the unit or hanging it up.

An experienced operator, using a medium to large size tractor, 80-Hp and up if crawler type, 160-Hp and up if rubber-tired, with ground conditions affording good traction and no problems regarding steep slopes, etc., should fell trees at the following rates:

Approx. Diam. of Tree	Approx. Height of Tree	Minutes Req'd. for Felling
8" to 12"	20'-0" to 25'-0"	5 to 10
15" to 24"	30'-0" to 50'-0"	15 to 20
27" to 36"	55'-0" to 75'-0"	25 to 30

Add for pushing felled trees to disposal area and labor disposing of same.

02110   DEMOLITION

Operating Costs on Compressed Air Units.—To reduce operating and overhead costs to a minimum, two tools should be operated from each compressor, otherwise the fixed charges, such as depreciation, interest, repairs, etc., will run rather high per unit.

The costs given on the following pages are based on using

two tools with each compressor unit. It should be borne in mind that the performance data is approximate only as job conditions may cause considerable variation in hourly or daily output. The wage rates quoted are used only as a guide in working out various examples. Insert wage rates applicable in your locality.

Tables showing air requirements for tools are given in the preceding chapter.

Breaking Pavements Using Pneumatic Concrete Breakers.—Costs on this work will vary with the hardness and thickness of the concrete, but on a pavement consisting of 6" concrete and 1½" to 2" of asphalt, 2 men operating concrete breakers and one compressor operator should break up 450 to 550 sq. ft. of pavement per 8-hr. day, at the following cost per 100 sq. ft.:

	Hours	Rate	Total	Rate	Total
Jackhammermen...	3.2	$....	$....	$12.54	$40.13
Compressor operator.............	1.6	....	....	17.05*	27.28
Compressor expense...............	1.6	....	....	13.00	20.80
Cost per 100 sq. ft.........................			$....		$88.21
Cost per sq. ft.......			....		.88

*Add only if required.

Cutting Out Concrete Foundations and Retaining Walls Using Pneumatic Concrete Breakers.—It is little more difficult to estimate the cost of cutting out heavy concrete walls, as so much depends upon the hardness of the concrete and the amount of reinforcing steel. However, on unreinforced 1:3:5 concrete such as used for ordinary foundations, 2 jackhammermen should cut out 225 to 275 cu. ft. of concrete per 8-hr. day, at the following cost per 100 cu. ft.:

	Hours	Rate	Total	Rate	Total
Jackhammermen...	6.4	$....	$....	$12.54	$80.26
Compressor operator.............	3.2	....	....	17.05*	54.56
Compressor expense...............	3.2	....	....	13.00	41.60
Cost per 100 cu. ft.........................			$....		$176.42
Cost per cu. ft.......			....		

*Add only if required.

02200 EARTHWORK

Estimating Quantities of Excavation.—Excavating is measured by the cu. yd. containing 27 cu. ft.

General or Mass Excavation.—The amount of slope necessary to provide a "safe" hole will vary with the kind of soil to be excavated, depth of excavation, etc. Digging in previously undisturbed material and assuming that no water or unstable conditions exist most estimators, when taking off quantities for excavation, use a 1 : 1 slope, that is one horizontal to one vertical, for sand and gravel; a 1 : 2 slope for ordinary clay and a 1 : 3 or 1 : 4 slope for stiff clay.

In some cases job conditions may not permit the sloping of banks and then it is usually necessary to sheet and brace the banks to prevent accidents from occurring resulting in damage to adjoining property and injury to workmen on the job.

Trench and Pit Excavation.—This kind of excavation is always estimated separately from the general work as it is usually a more expensive operation.

In many instances a good portion of this work may be done with power equipment, with some handwork required to clean out corners and grade the bottoms. On the other hand there may be portions of the work which must be done entirely by hand and which would be much more costly than machine work. In view of the wide variance in unit cost between machine work and handwork, the estimator should decide, when he is taking off the quantities, into which category each portion of the work will fall and list the machine work separately from the handwork so that each may be properly priced.

Quantities for this work are measured and listed in the same manner as for general excavation, except the slope factor may usually be ignored, unless the additional depth exceeds 4'-0" or if the ground will not stand safely for this depth with vertical banks. Under these conditions, allowances for sloping the banks should be made in the same manner as for general excavation.

Backfill.—After all excavating items have been figured the amount of backfill should be computed.

An easy method of doing this is by computing the displacement volume of the construction which is to be built within the limits of the excavation, that is, the volume of footings, piers, the basement volume figured from the underside of the fill under the floor to the elevation of the top of the general cut, etc., and deducting this volume from the sum of the general or mass excavation and the trench and pit excavation. The remainder is the volume of backfill required.

Another method of figuring backfill, usually more difficult than the one previously described, is that of computing the volume from the actual dimensions of the spaces requiring backfill.

Some specifications require interior backfills to be made with sand, gravel, bankrun gravel, etc., while the exterior backfill may be excavated material. In this case it is necessary to keep the two items separate as the interior backfill material will probably have to be purchased and brought in from outside sources.

Backfill very often is specified to be compacted by mechanical tampers and a degree of compaction is also specified. In these instances the backfill operations must be considered very carefully as this is expensive work.

Disposal of Excess or Borrow.—After all excavation, backfill, grading cuts and fills, etc., have been computed and listed in the estimate, the total cuts and the total fills should be compared to determine whether there will be an excess of materials to dispose of or a deficiency of materials to be made up by borrowing or purchasing from outside sources and the results of this comparison should be listed in the estimate. The top soil comparison should be kept separate from the other materials because the cost of any top soil which must be purchased usually is much more expensive than ordinary fill material. The difference between the cuts and fills will give the bank or compacted volume required. In the case of a deficiency of materials this compacted volume quantity must be adjusted for shrinkage due to compaction as the material will probably be bought by loose measure. In the same manner all excess bank measurement volumes must be adjusted for swell due to bulking to obtain the true volume which must be handled in the disposal operation. The amount of increase for gravels and sand is from 5 to 12 percent and for clays and loams from 10 to 30 percent.

Pumping or De-watering.—Finally, careful consideration should be given to the probability of pumping or de-watering operations. In some parts of the country this is no problem whatsoever, quite the reverse being the case. However, in most localities ground or climatic conditions make it necessary to include pumping or de-watering as an item in the estimate.

This item is quite variable, being affected by the season of the year as well as the locale. If only rainwater run-off is expected, an allowance for a few small pumps may be sufficient. If, however, the job is in close proximity to a body of water or springs are present, etc., it may be necessary to de-water part or all of the operational area by means of a well point system in which case the contractor should consult a company which specializes in this work.

METHODS OF EXCAVATING

Excavating costs will vary with the kind of soil and the method used in loosening and removing same.

Hand Excavating Production and Quantities

Sand or Loam	Ave. No. Cu. Yds 8-Hr. Day	Labor Hours Per Cu. Yd.
Shoveling and loading trucks by hand	8 to 10	0.90
Trenches to 5'-6" Deep	7 to 8	1.07
Pits and piers to 6'-0" Deep (1 lift)	6.	1.33

Sand or Loam	Ave. No. Cu. Yds 8-Hr. Day	Labor Hours Per Cu. Yd.
Pits and piers to 12'-0" Deep (2 lifts)	6	2.67
Pits and piers to 18'-0" Deep (3 lifts)	6	4.00
Picking, loosening and shoveling frozen sand or loam	1 to 2	5.33
Backfilling by hand, no tamping	18 to 20	0.40
Spreading loose sand or loam by hand	35 to 40	0.21
Ordinary Soil		
Shoveling and loading trucks by hand	5 to 6	1.375
Trenches 5'-6" Deep	4 to 5	1.75
Pits and piers to 6'-0" Deep (1 lift)	4	2.00
Pits and piers 12'-0" Deep (2 lifts)	4	4.00
Pits and piers to 18'-0" deep (3 lifts)	4	6.00
Backfilling by hand, no tamping	16 to 18	0.47
Spreading loose soil by hand	30 to 35	0.25
Heavy Soil And Clay		
Shoveling and loading trucks by hand	3½ to 4½	2.00
Trenches 5'-6" Deep	3 to 4	2.25
Pits and piers to 6'-0" Deep (1 lift)	3 to 4	2.25
Pits and piers to 12'-0" Deep (2 lifts)	3 to 4	4.50
Pits and piers to 18'-0" Deep (3 lifts)	3 to 4	6.75
Backfilling by hand, no tamping	12 to 14	0.617
Spreading loose clay by hand	23 to 27	0.32

Excavating Using a 1 Cu. Yd. Tractor Shovel

Class of Work	Ave. No. Cu. Yds. per 8-Hr. Day	Tractor and Operator Hours Per 100 Cu. Yds.
Excavating ordinary soil and placing in piles on premises	325 to 375	2.30
Excavating heavy soil and clay and placing in piles on premises	250 to 300	2.90
Loading loose sand or gravel into trucks	400 to 450	1.90

Excavating ordinary soil and loading into trucks	400 to 450	1.90
Excavating heavy soil and clay and loading into trucks..............	300 to 350	2.50
Backfilling loose earth—Bulldozing........................	500 to 600	1.45

The above table is based upon a 50-foot haul from excavation to dump. For each additional 50 feet in length of haul, add 1 ½ to 2 hrs. tractor time per 100 cu. yds. or consider the use of trucks.

Excavating Using a 2¼ Cu. Yd. Tractor Shovel

Class of Work	Ave. No. Cu. Yds. per 8-Hr. Day	Tractor and Operator Hours Per 100 Cu. Yds.
Excavating ordinary soil and placing in piles on premises.......	600 to 650	1.30
Excavating heavy soil and clay and placing in piles on premises	450 to 500	1.70
Loading loose sand or gravel into trucks...	775 to 825	1.00
Excavating ordinary soil and loading into trucks.......................	775 to 825	1.00
Excavating heavy soil and clay and loading into trucks..............	575 to 625	1.35
Backfilling loose earth—bulldozing	925 to 975	0.85

The above table is based upon a 50-foot haul from excavation to dump. For each additional 50 feet in length of haul, add ¾ to 1 hrs. tractor time per 100 cu. yds. or consider the use of trucks.

Approximate Rental Costs For Excavating Equipment With Operator

Type	Size	Day	Week	Month
Backhoe	½ cu. yd.	$ 370	$1,000	$ 4,200
	1 " "	600	1,700	6,900
	2 " "	900	4,000	16,000
Compactor*	1000# Ram.	47	225	800
	1000# Vib.	39	160	750
	5000# Vib.	180	750	2,800
	2000# Drum	80	350	1,300
Crane-Cable Crawler	30 T	560	2,500	9,500
	40 T	640	2,900	11,800
	60 T	720	3,200	13,100
Crane-Cable Truck	40 T	760	3,300	14,000
	60 T	900	3,900	16,400
	80 T	1,100	4,800	19,600

Type	Size	Day	Week	Month
Crane Hydro Truck	5 T	470	2,100	8,400
	15 T	525	2,300	9,500
	30 T	570	2,500	10,300
Drill*	40# Rock	25	75	200
	65# Rock	30	85	240
Scraper	11 cu. yd.	500	2,250	9,000
	22 " "	850	3,900	15,800
Tractor - Dozer	105 HP	400	1,900	7,600
	180 "	450	2,100	8,600
Loader	1½ cu. yd.	400	1,900	7,600
	2¼ " "	480	2,300	9,100
Trencher	16 "	330	1,400	4,500
	24 "	400	1,600	5,800

*No Operator

The above costs do not include move charges or permits

Power Shovel Yardages.—To estimate yardage production on a job is a real problem. If all conditions were the same on all jobs, it would be a simple matter, but jobs are as different as night and day. There are many factors which must be considered, many of which can be learned only through experience.

The type of material, the depth of cut for maximum effect, no delays in operation, a 90° swing for the shovel, all affect the output, let alone the conditions existing on the job.

The following table gives an approximate idea of the difference in maximum output possible, subject to the conditions as listed above.

Hourly Shovel Output in Cubic Yards

Class of Material	3/8	1/2	3/4	1	1¼	1½	1¾	2	2½
Moist Loam or Light Sandy Clay	85	115	165	205	250	285	320	355	405
Sand and Gravel.	80	110	155	200	230	270	300	330	390
Common Earth ... Clay,	70	95	135	175	210	240	270	300	350
Hard and Tough. Rock, Well Blasted	50	75	110	145	180	210	235	265	310
Common Earth, with	40	60	95	125	155	180	205	230	275
Rocks and Roots. Clay,	30	50	80	105	130	155	180	200	245
Wet and Sticky... Rock,	25	40	70	95	120	145	165	185	230
Poorly Blasted.....	15	25	50	75	95	115	140	160	195

(Shovel Capacity)

The quantities given in the above table are based on bank measure which means cubic yards removed from the bank rather than cubic yards in the hauling unit. There is a big difference between a yard of dirt in the bank and a yard of dirt in the truck. This is because of the swell of the material or its increase in volume due to voids when it is dug or loosened. For instance, common excavation will swell from 10 to 30 percent. Another condition is that the machine is working in a depth of cut suitable for maximum digging efficiency. The optimum depth of cut for various sizes of shovels, may be defined as that depth which produces the greatest output and at which the dipper comes up with a full load.

The output figures are based on continuous operations, 60 minutes a working hour, without any delays for adjustments, lubrication, operator stopping for any reason, etc.

Dragline Yardages.—As in the case of power shovels and hoes, dragline yardage production is affected by the type of material to be excavated, the depth of cut, the swing before unloading, the type of unloading (unloaded onto hauling units, cast onto spoil banks, etc.), the degree of continuity in the operation, etc.

The following table gives an approximate idea of the difference in maximum production possible subject to conditions listed thereafter:

Hourly Short Boom Dragline Output in Cubic Yards

| Class of Material | Bucket Capacity | | | | | | | | |
	³⁄₈	½	¾	1	1¼	1½	1¾	2	2½
Moist Loam or Light Sandy Clay	70	95	130	160	195	220	245	265	305
Sand and Gravel	65	90	125	155	185	210	235	255	295
Common Earth	55	75	105	135	165	190	210	230	265
Clay, Hard and Tough	35	55	90	110	135	160	180	195	230
Clay, Wet and Sticky	20	30	55	75	95	110	130	145	175

The dragline is working in the optimum depth of cut for maximum efficiency.

The dragline is working a full 60 minutes each hour—no delays

The dragline is making a 90° swing before unloading.

The bucket loads are being dumped into "properly sized" hauling units.

The proper type bucket is being used for the job.

Clamshell Production.—The clamshell excavator is not to be considered a high production machine but rather a machine to be used where the work is beyond the scope of other types of equipment. For example, a condition which usually requires a clamshell is where digging is vertical or practically straight down as in digging pier holes, shafts, etc. Digging in trenches which are sheathed and cross-braced generally calls for a clamshell because the vertical action of the bucket enables it to be worked through the cross-bracing. Jobs requiring accurate dumping or disposal of materials are usually clamshell jobs. Also, for high dumping jobs, whether it be charging a bin, building a stockpile or wherever the material must be dumped well above the machine level, the machine is well adapted. In general, the clamshell can operate vertically and dig or spot dump below, at or above the level of the machine.

The clamshell is only effective where the materials to be handled are relatively soft or loose.

Hourly Short Boom Clamshell Handling Capacity in Cubic Yards

Bucket Capacity	3/8	1/2	3/4	1	1 1/4	1 1/2	1 3/4	2	2 1/2
Class of Material									
Moist Loam or Sandy Clay........	50	65	95	120	140	155	170	190	225
Sand and Gravel..	45	60	85	110	130	140	160	175	205
Common Earth	40	55	70	95	115	125	145	160	185

It must be thoroughly understood that the above production is based on the most ideal conditions with 100% job and management factors.

Cost of Hauling Excavated Material in Trucks.—Hauling costs vary with the capacities of the trucks and time of the hauling cycle.

Truck capacity, for proper sizing, should be at least 4 times the dipper or bucket capacity of the excavator. This is important as the efficiency of the excavator is seriously affected by undersized hauling units due to the increased hazards of truck delays.

The following table gives theoretical spotting time cycles for various sizes of power shovel excavators working at 100 percent efficiency and loading trucks of 4 times dipper size capacity.

For dragline operation the above time values may be increased from 20 to 30 percent.

The hauling time cycle consists of several operations, namely; spotting truck under excavator; loading truck by excavator; traveling to dumping area; dumping; returning to excavator.

Some of these operations, such as spotting, loading and dumping, can be minimized by good management and supervision. Traveling time, however, will vary with the haul distance, traffic conditions, road conditions, etc., and must be carefully analyzed for each job.

When the hauling time cycle has been estimated, the number of loads each truck is able to haul per day can be determined and multiplying this by the capacity of the truck will give the theoretical daily haul of each truck in cu. yds., assuming 100 percent job efficiency. This amount should be multiplied by the expected job efficiency factor to obtain the net yardage hauled. Divide the daily truck cost by the net yardage hauled and the result will be the hauling cost per cu. yd.

Tractors and Scrapers

Stripping and stockpiling top soil on small quantity jobs with a haul distance from 200 to 400 feet or less is most efficiently

Haul Units Needed to Spot Under Shovel per Hour in Medium Digging

Size Excavator Dipper	Minimum Haul Unit Capacity at 4 Times Dipper Size	Approximate Shovel Cycle in Seconds 90° Swing No Delays	Loading Time for 4-Dipper Truck in Seconds	Time of Spotting Cycle for Steady Operation in Minutes	Number of Spots Required per Hour for Steady Operation
⅜	1½ Yd.	19	76	1.26	48
½	2 Yd.	19	76	1.26	48
¾	3 Yd.	20	80	1.33	45
1	4 Yd.	21	84	1.40	43
1¼	5 Yd.	21	84	1.40	43
1½	6 Yd.	23	92	1.53	39
2	8 Yd.	25	100	1.66	36
2½	10 Yd.	26	104	1.73	35

done by bulldozer. On larger jobs this is done more economically by scrapers.

Scrapers are high production earth moving units which are capable of digging their own load, hauling the load and then spreading the load in controlled layers. Furthermore, the scraper is a precision tool, as a skilled operator can cut a grade to within 0.1 ft. and can spread fill to the same degree of accuracy.

Scrapers may be powered by three types of prime movers: two wheeled or four wheeled rubber-tired units or crawler type tractors.

The basic efficiency of scrapers is relatively unaffected by depth of cut, length of haul or type of soil. Quantities being sufficient to warrant the use of scraper equipment, crawler tractor drawn scrapers are economical at haul distances up to 500 feet and with rubber-tired, high speed prime movers can compete with trucks on longer hauls.

Capacities of scrapers range from about 7 cu. yds. struck measure to approximately 40 cu. yds. heaped measure.

Scrapers powered by crawler tractors are capable of self-loading in most soils but generally it is more economical to use pusher assistance from a tractor to obtain heaped loads in the shortest time. With few exceptions, pusher assistance is essential for self-propelled, rubber-tired scrapers. Pusher loading is generally accomplished in an average of about 1 minute in 100 ft. of travel. To determine the number of scrapers one pusher can handle, divide the scraper cycle by the pusher cycle.

The following table gives the daily scraper production for various sized machines operating at several different time cycles and is based on ordinary soil conditions, good weather, good equipment and an 83 percent job efficiency.

Daily Scraper Production for Various Operation Cycles
Based on a 50-Minute Hour and an 8-Hour Day

Operation Cycle in Minutes	Struck Capacity of Scraper in Cu. Yds.		
	7	14	24
3	930	1860	3200
4	700	1400	2400
5	560	1120	1920
6	465	930	1600
7	400	800	1370
8	350	700	1200
9	310	620	1070
10	280	560	960

Example Estimate for Scraper Excavation

Based on a site grading cut containing 8,000 cu. yds., excavated material to be spread-dumped in low areas requiring fill, good conditions permitting a 50-minute hour, an operation cycle of 6 minutes and using two 14 cu. yd., struck measure, self-propelled scrapers.

	Hours	Rate	Total	Rate	Total
Move-in charge					
2 ea....................		$....	$....	$125.00	$ 250.00
Excavating					
Foreman................	35	....	....	18.45	645.75
Scraper operators .	70	....	....	17.95	1,256.50
Scraper Hours.......	70	....	....	91.31	6,391.70
Total Direct					
Cost....................			$....		$8,543.95
Cost per cu. yd......			....		1.06

If a 105-Hp. bulldozer would be required to give pusher assistance in loading, to maintain the operation cycle time of 6 minutes, the additional cost per cu. yd. would be as follows:

	Hours	Rate	Total	Rate	Total
Move-in charge		$....	$....	$100.00	$ 100.00
Excavating					
Bulldozer operator	35	....	....	17.95	628.25
Bulldozer charge...	35	....	....	30.66	1,073.10
Additional					
Direct Cost.........			$....		$1,801.35
Additional cost					
per cu. yd.			....		.23

Trenching and Ditching Machine Excavation

Designed strictly for the one purpose of cutting ditches, the trenching machine, when properly applied, is the fastest earthmover for its weight and horsepower in the construction field. Before a contractor can reap the benefits of this great potential output, he must first select a trenching machine of a type and size which fits the work to be done and, even more important, must put an operator on the machine who possesses both skill and mechanical knowledge.

There are two basic types of machines, the wheel type and the ladder type. The wheel type is generally considered to be the fastest and is by far the most common type in use today. Each type has advantages and disadvantages for performing certain classes of work.

Inasmuch as most trenching or ditching operations in connection with building construction will be foundations, service trenches, etc., the following data covers the cost and operation of a general purpose, crawler mounted, wheel machine of the type and size generally used for this work.

This machine is equipped with shifting-tilting boom that permits it to dig close to obstructions, such as foundations or curbs, and to dig a vertical trench even when the machine is on a slope.

The cost of excavating using a trenching or ditching machine will vary with the size of the job, width and depth of trenches, kind of soil, accessibility, etc. Most machines are designed to dig certain standard width trenches, so that trenches a trifle narrower or wider will not make any appreciable difference in either output or costs.

While the following table gives the rated capacity of the machine, it will be necessary to allow a factor for time lost moving machine, bad weather, etc., so it is not advisable to figure rated capacity as the average daily output of any machine.

Data on Crawler Mounted Ditching Machine Operation and Cost

Wheel Size	Possible Cutting Widths	Maximum Depth Trench	Weight in Pounds	Approx. Cost F.O.B. Fcty.
16"x11'-6"	16"-24"	7'-0"	27,000	$130,000
24"x12'-4"	24"-30"	7'-6"	43,000	$170,000

Maximum digging speed is 34 f.p.m. Forward and reverse travel speeds are 0.46, 0.95, 1.7 and 2.9 m.p.h.

Digging speeds are controlled by a hydraulic transmission, permitting speed to be varied to match soil conditions. Wheel speeds are variable from 0 to 9.2 r.p.m.

Cost of Digging 100 Lin. Ft. of Trench 5'-0" Deep, Using a Crawler Mounted Ditching Machine, Based on 350 Lin. Ft. per Hour

	Hours	Rate	Total	Rate	Total
Machine operator	0.29	$....	$....	$17.95	$ 5.21
Machine charge, 12" wide		....	....	.193	19.30
Machine charge, 24" wide		....	....	.248	24.80
Cost per 100 lin. ft., 12" wide					24.51
Cost per 100 lin. ft., 24" wide					30.01
Cost per lin. ft., 12" wide					.25
Cost per lin. ft., 24" wide					.30

To the above, add cost of trucking machine to job and removing same at completion, supervision, compensation and liability insurance, Social Security and Unemployment taxes, overhead expense and profit. Add for backfilling trenches, if required.

Grading

Rough Grading.—After all backfill is in place and the site has been cut and/or filled to the approximate specified contours, rough grading of the site is done, by hand for small jobs and by machine for large areas, usually preparatory to the spreading of top soil. Tolerance for this type grading is usually plus or minus 0.1 ft.

On small jobs a laborer should rough grade 800 sq. ft. of ground surface per 8-hr. day at the following cost per 100 sq. ft.:

	Hours	Rate	Total	Rate	Total
Labor	1.0	$....	$....	$12.54	$12.54
Cost Per Sq. Ft.			....		.13

On large jobs a bulldozer is usually employed to do rough grading and a laborer generally accompanies the machine to check the surface and direct the operation. A 75-hp. bulldozer should rough grade 6,500 to 7,500 sq. ft. per 8-hr. day at the following cost per 1,000 sq. ft. :

	Hours	Rate	Total	Rate	Total
75-Hp. bulldozer charge	1.15	$....	$....	$23.80	$27.37
Bulldozer operator	1.15	....	....	17.95	20.64
Labor	1.15	....	....	12.54	14.42
Cost per 1,000 sq. ft.			$....		$62.43
Cost per sq. ft.			....		.06

Does not include hauling equipment to and from job, compensation and liability insurance, Soc. Sec. taxes or overhead and profit.

Grading for Slabs on Ground.—Grading for slabs on ground such as floors, walks, driveways, etc., usually is done by hand unless the job is quite large and can be organized so that some of the work is done by machine. This work generally must be done accurately with tolerances of no more than ½" allowed. A laborer should grade 500 to 600 sq. ft. per 8-hr. day at the following cost per 100 sq. ft. :

	Hours	Rate	Total	Rate	Total
Labor	1.5	$....	$....	$12.54	$18.81
Cost per sq. ft.			....		.19

For sloping surfaces add about 50 percent to the above cost depending on steepness of pitch.

Finish Grading of Top Soil.—After top soil has been spread over areas specified, a finish grading operation, usually including hand-raking, must be performed prior to seeding or sodding. The tolerance on this work usually is plus or minus 1 inch. A laborer should finish grade and hand-rake 600 to 700 sq. ft. per 8-hr. day at the following cost per 100 sq. ft. :

	Hours	Rate	Total	Rate	Total
Labor	1.25	$....	$....	$12.54	$15.68
Cost per sq. ft.			....		.16

For sloping surfaces, such as berms, ditch sides, etc., add 50 percent to the above cost.

Grading for Footing Bottoms.—When footing pits and trenches are dug by machine there is always cleanup, squaring and grading work to be done by hand and many contractors price this work by the sq. ft. of footing bottom. A laborer should clean up, square and grade 200 sq. ft. of footing bottom per 8-hr. day at the following cost per 100 sq. ft. :

	Hours	Rate	Total	Rate	Total
Labor	4.0	$....	$....	$12.54	$50.16
Cost per sq. ft.			....		.50

Costs Of Excavating

Loader Excavating under Favorable Conditions.—The area of loose clay and loam is 80'-0"x80'-0"x5'-0", a total of 1,185 cu. yds. of excavation.

The average on this job is 135 cu. yds. an hr. at the following cost per 100 cu. yds.:

	Hours	Rate	Total	Rate	Total
Loader and Operator	.75	$....	$....	$50.00	$37.50
Cost per cu. yd.			....		.38

Five 6-cu. yd. trucks will haul 1,080 cu. yds. of excavated earth ¼-mile to a dump per day, or an average of 27 cu. yds. an hr. per truck, at the following cost per 100 cu. yds. :

	Hours	Rate	Total	Rate	Total
Truck and Driver	3.7	$....	$....	$35.00	$129.63
Cost per cu. yd.			....		1.30

Excavating Using Pneumatic Diggers.—These tools can be used to advantage on all classes of excavation that requires picking of dirt, clay, etc., such as trench work, tunneling, caisson sinking and all kinds of building excavation; in fact, on all kinds of work in stiff clay or hard ground where power shovels, trenching machines, etc., cannot be used.

On work of this kind, two men using air picks and six men shoveling will loosen and shovel 35 to 40 cu. yds. per 8-hr. day, at the following labor cost per cu. yd. :

	Hours	Rate	Total	Rate	Total
Labor operating picks	0.4	$....	$....	$12.54	$ 5.02
Labor shoveling	1.2	....		12.54	15.05
Compressor operator	0.2	....		17.05*	3.41
Compressor expense	0.2	....		13.00	2.60
Cost per cu. yd.			$....		$26.08

*Add only if required.

On larger excavations providing plenty of room for the men to work, a man with pneumatic pick will loosen 25 to 30 cu. yds. per 8-hr. day. After the clay or earth is loosened, a man will shovel 6 to 7 cu. yds. per 8-hr. day.

Comparative Cost By Hand

Excavating in stiff clay or tough ground, a man using a hand pick and shovel will loosen 2½ to 3 cu. yds. of dirt per 8-hr. day. The same man, picking only, will loosen 3½ to 4 cu. yds. of dirt per 8-hr. day, and when shoveling only, will remove 6 to 7 cu. yds. per 8-hr. day.

Using a hand pick, a man will loosen and shovel one cu. yd. of dirt at the following cost per cu. yd. :

	Hours	Rate	Total	Rate	Total
Labor	3	$....	$....	$12.54	$37.62

02250 SOIL TREATMENT

Often a soil poisoning will be made a part of the contract for termite control. Termites are found in most all states but are more prevalent in the south, southwest, southern Pacific, and Ohio Valley regions.

The usual chemicals encountered are chlordane, 1% emulsion, mixed 1 gal. of concentrate to 48 gals. of water; dieldrin, 0.5% emulsion mixed 1 to 36 gals. water; and benzene hexachloride, 0.8% emulsion mixed 1 to 15 gals. of water. There are oil mixtures, but water will not stain or creep, nor does it damage foundation plantings.

The rate of application is usually 2 gals. of diluted emulsion per 5 lin. ft. of wall per each 12 in. of depth treated. The inside of a crawl space or on grade slab is usually trenched only 6" to 8" deep and the same width. The emulsion is poured in, then backfilled, and another application made for a total of 4 gals. per each 5 lin. ft. of wall. Remember to include piers also. Applications on exterior of walls is made at the same rate, a new application being made after each 12 in. of depth of backfill. A 3 foot deep foundation wall would thus require a 6 gal. emulsion per each 5 ft., while a 6 ft. deep basement wall would require twice that.

02300 PILE FOUNDATIONS

Setting up and Removing Equipment.—The cost of moving the pile driving equipment to the job must be taken into consideration, and then after it arrives at the job, there is the task of setting up the pile driver and getting ready to operate.

After the pile driver has been delivered at the job it should take about 3 days time for the crew to set it up ready to drive and this labor should cost as follows :

Setting up Equipment	Hours	Rate	Total	Rate	Total
Foreman	24	$....	$....	$18.62	$ 446.88
Engineer on pile driver	24	....	....	17.95	430.80
Oiler	24	....	....	12.54	300.96

Setting up					
Equipment	Hours	Rate	Total	Rate	Total
2 pile driver men..	48	....	....	16.47	790.56
4 men, general labor	96	....	....	12.54	1,203.84
Total labor cost.			$....		$3,173.04

The above item will remain practically the same whether there are 100 or 1,000 piles to be driven.

After the job has been completed, the pile driver must be dismantled and removed from the job and the dismantling should cost as follows :

Dismantling					
Equipment	Hours	Rate	Total	Rate	Total
Foreman	16	$....	$....	$18.62	$ 297.92
Engineer on pile driver	16	....	....	17.95	287.20
Oiler	16	....	....	12.54	200.64
2 pile driver men	32	....	....	16.47	527.04
4 men, general labor	64	....	....	12.54	802.56
Total Labor Cost..			$....		$2,115.36

WOOD PILES

Driving Wood Piles.—After the pile driver is set up and ready to operate, the labor cost of driving the piles will vary with the length of the piles and job conditions.

The following table gives the approximate number of piles of various lengths that should be driven per hour and per day :

Length of Piles	1 Hour	8 Hours
24 feet	4.5	36
26 feet	4.2	34
28 feet	4.0	32
30 feet	3.7	29
32 feet	3.5	28
34 feet	3.3	26
36 feet	3.1	25
38 feet	3.0	24
40 feet	2.8	22
45 feet	2.5	20
50 feet	2.3	18
55 feet	2.1	17
60 feet	2.0	16

The labor operating the pile driver per 8 hr. day should cost as follows :

	Hours	Rate	Total	Rate	Total
Foreman	8	$....	$....	$18.62	$ 148.96
Engineer on pile driver	8	....	....	17.95	143.60
Oiler	8	....	....	12.54	100.32

2 pile driver men ..	16	....	....	16.47	263.52
4 men, general labor..................	32	....	....	12.54	401.28
Labor cost per day			$....		$1,057.68

The above does not include time laying out the work or spotting piles, pumping, shoring, excavating or cutting off wood piles after they are driven, but includes only the actual time driving the piles. Neither does it include the cost of equipment, fuel, oil, and other supplies for the pile driver, which should be added.

Cutting off Wood Piles.—After the wood piles have been driven, the tops projecting above the ground will have to be cut off to receive the foundation which is to be placed upon them.

This cost will vary according to the size of the quarters in which the men are obliged to work, but on the average job it should require 1/6 to 1/4 hrs. to cut off each wood pile 12" or 14" in diameter with a chain saw, at the following labor cost per pile :

	Hours	Rate	Total	Rate	Total
Labor	0.22	$....	$....	$16.47	$3.62

To the above costs, add the hourly cost of chain saw, and the time for removing sawed-off ends of the piles from the premises.

Computing the Cost of Wood Piles.—As described above, the cost of driving wood piles will vary with the number of piles, length, kind of soil, and other job conditions. Costs of the piles fluctuate so rapidly that current costs must be verified for each project.

An example of an estimate showing a method for arriving at costs for driving 680 treated wood piles 40'-0" long is as follows:

	Local Rate	Total	Total
Trucking pile driver to job ...	$....	$....	$ 1,500.00
Setting up pile driver, 3 days for crew	....	....	3,173.04
680 wood piles, 40'-0" long @ $160.00	....	....	108,800.00
Driving piles (22 per 8-hr. day) 31 days @ $1,057.68 .	....	....	32,788.08
Fuel, oil, misc. supplies, etc. 31 days @ $150.00	....	....	4,650.00
Dismantling pile driver at completion, 2 days..................	....	....	2,115.36
Trucking pile driver, job to yard...................................	....	....	1,500.00
Equipment rental or depreciation allowance (36 days)....................................	....	....	18,000.00
Cost 680 piles.......................		$....	$172,526.48

	Local Rate	Total	Total
Cost per pile............................		....	253.72
Cost per lin. ft.		....	6.34

Add cost of cutting off piles after driving, and for overhead & profit.

CONCRETE PILES

There are two principal types of concrete piles, cast-in-place and precast. The cast-in-place pile is formed in the ground in the position in which it is to be used in the foundation. The precast pile is cast above ground and, after it has been properly cured, it is driven or jetted just like a wood pile.

In ordinary building foundation work the cast-in-place pile is more commonly used than the precast pile for the following reasons. For cast-in-place piles, the required length can readily be adjusted in the field as the job progresses. Thus there is no need to predetermine pile lengths and the required length is installed at each pile location.

In marine installations either in salt or in fresh water, the precast pile is used almost exclusively, because of the difficulty involved in placing cast-in-place piles in open water. For docks and bulkheads, the cast-in-place pile is sometimes used in the anchorage system. On trestle type structures such as highway viaducts, the precast pile is more commonly used. A portion of the pile often extends above the ground and serves as a column for the superstructure.

Raymond Concrete Piles*—The basic Raymond piles are Step-Taper and Pipe Step-Taper. They are steel-encased cast-in-place concrete piles installed by :
1) placing a steel shell closed at the tip over a steel mandrel,
2) driving the mandrel and shell to the required resistance and/or penetration,
3) withdrawing the mandrel, leaving the steel shell in place as driven,
4) inspecting the steel shell internally its full length,
5) removing excess shell, and
6) filling the steel shell with concrete to pile cut-off grade.

The steel shell not only maintains the driving resistance but also protects the fresh concrete from being mixed with the surrounding soil or from being distorted by soil pressures set up by the driving of adjacent piles.

Raymond Step-Taper Pile* shells are made of sheet steel in gages of 12 to 20 and in basic section lengths of 4, 8, 12 and 16 feet. Longer lengths can be furnished for special conditions. Step-Taper shells are helically corrugated to provide greater strength against collapsing pressures. A driving ring is welded to the bottom of each shell section and joints between sections are screw-connected. The bottom section is closed with a flat steel plate welded to the driving ring. Step-Taper Pile shells are manufactured in standard nominal diameters ranging from

8 5/8 inches to 18 3/8 inches but larger diameters can be made to meet special conditions.

When sufficient shell sections are joined together to make up the required pile length, the pile diameter increases from tip to butt at the rate of one inch per section length. Thus the rate of taper or pile shape will vary with the section lengths used. Within practical limits, different section lengths can be combined in a single pile thus providing a wide range of available pile shapes. Nominal tip diameters usually range from 8 to 11 inches but larger tip diameters can be used. With the variety of diameters and section lengths available, the Step-Taper pile provides maximum flexibility in the choice of pile size and configuration to best meet subsoil conditions and loading requirements.

The length of a Step-Taper Pile can be readily adjusted as the work proceeds, to meet variable soil conditions. Thus waste costs are held to a minimum. It is unnecessary to install drive-test piles to predetermine pile lengths.

Internal reinforcement is not required for Step-Taper Piles except to resist uplift loads or high lateral loads (where batter piles are not used) or for unsupported pile lengths extending through air, water or very fluid soil.

Step-Taper Piles are suitable for all types of soils and can function as friction or point bearing piles. In most cases, piles are supported through a combination of friction and direct bearing. In many soils, tapered piles usually develop higher capacity than piles of no taper.

Step-Taper Piles are driven with a heavy rigid steel mandrel which provides effective hard driving and the development of high capacity piles. The maximum practical length for an all-shell Step-Taper Pile is about 140 feet. However, longer all-shell piles could be installed providing a pile driving rig of adequate capacity were available. Where exceptionally long, high capacity piles are required, the Raymond Pipe Step-Taper Pile is used.

Raymond Pipe Step-Taper Piles* are fundamentally the same as the Step-Taper Piles except that a closed-end steel pipe is used for the lower portion of the pile. In some cases, the mandrel used to drive the shells can extend through the pipe to the pile tip. The joint between pipe and shell is either a water-tight slip joint, a drive sleeve joint or a welded joint and is designed to transfer the driving force from the mandrel to the top of the pipe. If the mandrel extends for the full length of the pipe, the driving force is also carried down to the pile tip.

The diameter of the Step-Taper shell section just above the pipe is determined by the diameter of the pipe used and is either approximately the same size as the pipe or about one inch larger in diameter. Pipe diameters normally range from 10 to 14 inches although larger pipe sizes have been used.

The required pipe wall thickness would depend upon driving conditions, the length of the pipe portion and whether or not it were driven with an internal mandrel. For hard driving, high penetration resistance, long pipe portions or non-mandrel driven pipe, a heavier wall pipe is required for effective driving.

Pipe Step-Taper Piles are suitable for all types of soils and function as friction or point bearing piles. Through variations in lengths of pipe and shell portions, sizes of pipe and sizes and section lengths for the Step-Taper shells, the Pipe Step-Taper Pile also provides a wide range of pile size and configuration to best meet subsoil conditions and loading requirements. Pile lengths can readily be adjusted in the field to meet variable subsoil conditions.

The length of the Pipe Step-Taper Pile is not limited by the capacity of the pile rig. The full length pile can be driven in one piece—pipe and shell together—or the pile can be installed in two or more stages—the pipe first, followed by the Step-Taper shells. For a very long pipe portion, the pipe can be driven in sections which are joined by sleeves or welding.

Steel Pipe Piles.—Another kind of cast-in-place pile with a permanent shell is the steel pipe pile, both open and closed end.

The closed end pipe pile is simply a piece of steel pipe, closed at the bottom with a heavy boot, then driven into the ground and filled with concrete. The uses and allowable loads for such piles are about the same as other types of cast-in-place piles with driven shells. On some jobs the pipe shell is driven all in one piece, but very often it is driven in sections which are welded together or fitted together with internal sleeves as the driving progresses. Adequate wall thickness of the pipe is necessary to develop the required stiffness and thus driveability of the pipe.

Open end steel pipe piles are usually driven to bearing on rock. Since the pipe is open at the bottom during driving, the interior fills with soil which must be removed before concreting. Cleaning is usually done with air and water jets, after which the pipe is re-driven to insure proper seating in the rock, remaining water is pumped out and the pipe pile is filled with concrete.

Open end pipe piles installed in this manner are usually allowed to carry relatively high working loads. They have been installed up to 24 inches in diameter and in lengths up to 200 feet.

Steel H-Beam Piles.—Rolled structural steel shapes are also used as bearing piles. The shape commonly used for this purpose is the H-beam. This type of pile has proved especially useful for trestle structures in which the pile extends above the ground and serves not only as a pile but also as a column.

Because of their small cross-sectional area, piles of this type can often be driven through dense soils to point bearing where it would be difficult to drive a pile of solid cross-section, such as a wood, cast-in-place, or precast concrete pile. This ability to penetrate dense soils easily works to a disadvantage in other soil conditions. Where piles are used to support loads by friction or where they are used primarily for compaction, a considerably longer H-beam is required to carry the same load that can be supported by a pile of less length but greater cross-sectional area.

Prices of steel H-beam piles vary considerably, mainly due to the material cost of the rolled steel sections. Steel sections commonly used for this purpose may range from an 8-inch H-section weighing 36 lbs. per lin. ft. to a 14-inch H-section weighing 117 lbs. per lin. ft.

Assuming a unit cost of .30 cents per lb., delivered to job site from warehouse, the H-beam material will cost from $10.80 to $35.10 per lin. ft. If job is in an isolated location, additional transportation charges must be figured.

Under average conditions, the cost of driving steel H-beam piles will vary from $3.00 to $4.00 per lin. ft., not including moving the equipment on and off the job.

02350 CAISSONS

Caissons may be dug either with straight shafts to the required depth, or they may be belled at their bottoms to provide additional bearing area. The majority of caissons are now being installed using various types of drill rigs but under certain conditions hand dug caissons may be used. The drilling machines are capable of not only drilling the shaft, but mechanically forming a bell as well. The cost of excavating varies with the diameter and depth of the caisson, the site conditions, and with the type of soil being excavated.

A typical job consisting of 100 caissons with 2'-0" shaft diameters, 5'-0" bells and 25' deep, might be estimated as follows, assuming normal clay digging. First, compute the total volume of excavation which in this case would be 418 cu. yds. To this figure add about 10% overbreak, making a total of 460 cu. yds. Assuming each drill rig can complete 5 caissons per day, the job would take 20 days. Allow approximately 10% extra time for weather and total of 23 days to complete the job.

Labor	Hours	Rate	Total	Rate	Total
Drill Rig Operator	184	$....	$....	$ 17.95	$ 3,302.80
Oiler	184	....	....	12.54	2,307.36
Caisson Laborer....	184	....	....	13.54	2,491.36
Laborer	184		....	12.54	2,307.36
					$10,408.88
Equipment					
Drill rig	23 days	$....	$....	$400.00	9,200.00
Material					
Concrete	460 cu. yds.	....	....	$ 55.00	25,300.00
					44,908.88
Cost per cu. yd. based on	418 cu. yds	....	$....	....	$ 107.44

These costs are for the complete caisson installation. Allowances must be added for reinforcing steel, additional labor based on union requirements, temporary or permanent casings if necessary, soil and concrete testing, surveying, disposal of excavated materials, and equipment move and set-up charges, if these items are included in the caisson contractor's contract, and overhead and profit.

02400 SHORING

Wood sheet piling and bracing is usually estimated by the square foot, taking the number of sq. ft. of bank or trench walls to be braced or sheet pile, plus an additional amount for penetration at the bottom of the excavation, and an allowance for extension above the top, and estimating the cost of the work at a certain price per sq. ft. for labor and lumber required.

Driving Sheet Piling by Compressed Air.—Pneumatic pile drivers are very effective for driving sheet piling. They are essentially a heavy paving breaker equipped with a special fronthead which is adjustable for driving 2" to 3" piling. These machines weigh about 125 lbs. and will drive wooden sheet piling in any ground which the piling will penetrate, such as sand, gravel or shale and any gradation between soils.

Its driving speed varies from 2'-0" per minute in hard clay or shale to 9'-0" per minute in sand or gravel.

Under average conditions, two men working together on one machine should drive 100 lin. ft. (60 to 79 sq. ft.) per hr. at the following labor cost per 100 lin. ft.:

	Hours	Rate	Total	Rate	Total
Jackhammerman...	1	$....	$....	$12.54	$12.54
Helper......................	1	....	....	12.54	12.54
Compressor expense & Tools	1	....	....	13.50	13.50
Cost per 100 lin. ft..........................			$....		$38.58
Cost per lin. ft			....		3.86

Add for compressor engineer if required.

STEEL SHEET PILING

Steel sheet piling is used for supporting the soil in large excavations, cofferdams, caissons, and deep piers or trenches where wood sheet piling is impractical. It is driven with a pile driver the same as wood or concrete piles.

There are a number of different types of steel sheet piling on the market but the following will furnish an idea of the sizes, weights, etc., commonly used.

Sizes and Weights of United States Steel Sheet Piling

Section Number	Width Inches	Web Thick. Inches	Wt. Lbs. per Lin. Ft.	Wt. Lbs. per Sq. Ft. Wall
MZ38	18"	3/8"	57.0	38.0
MZ32	21"	3/8"	56.0	32.0

MZ27	18"	3/8"	40.5	27.0
MP102	15"	1/2"	40.0	32.0
MP101	15"	3/8"	35.0	28.0
MP113	16"	1/2"	37.3	28.0
MP112	16"	3/8"	30.7	23.0
MP110	16"	31/64"	42.7	32.0
MP116	16"	3/8"	36.0	27.0
MP115	19"5/8	3/8"	36.0	22.0

02500 SITE DRAINAGE

Site drainage includes such items as subdrainage, foundation and underslab drainage, drainage structures, sanitary and storm drainage piping, and dewatering and wellpoints.

Subsoil drains are usually vitrified clay pipe, set with lowest point at the same elevation as the bottom of the footing, pitched a minimum of 6 in. in 100 lin. ft., laid ¼" apart and joint covered with 15# asphalt felt or #12 copper or aluminum mesh. 4" pipe is minimum, 6" preferred. One laborer can lay approximately 10 ft. of 4" pipe per hour, 8 ft. of 6". 4" pipe costs around $1.00 per ft; 6", $1.40. Excavation is, of course, extra and it is assumed the trench is properly sloped with a firm bottom before pipe is installed.

All sharp turns shall be formed with fittings which cost $2.50 for 4" elbows, $4.75 for 6". Wide angles can be made by bevelling tile ends into easy radius bends. Catch basins can be built of masonry, poured with concrete, or be ordered precast.

For other than sewers, concrete, asbestos cement and corrugated metal pipe are often used. The relative material costs per lineal foot are approximately as follows:

Size of Pipe	Concrete	Asbestos Cement	Corrugated Metal
6"	$ 3.00	$ 2.50	$. . .
8"	3.25	3.00	3.70
10"	3.75	4.25	4.30
12"	4.00	6.00	5.25
18"	7.00	11.50	7.35
24"	11.00	19.00	12.50
30"*	15.00	27.00	15.00
36"*	22.00	36.00	23.00

*Reinforced

WELLPOINT SYSTEM OF DE-WATERING

A wellpoint system as used in construction work, usually consists of a series of properly sized wellpoints, surrounding or paralleling the area to be de-watered, connected to a header pipe by means of risers and swing piping. Header piping, in turn, is connected to one or more centrifugal pumps depending on the volume of water which must be handled.

Wellpoints are usually self-jetted into place to the correct depth and at the proper spacing to meet the requirements.

Lumber Required for Sheet Piling

Size of Timbers	Width Sq. Edge	100 Lin. Ft. Covers Sq. Ft.	Ft. B. M. per 100 Sq. Ft.	Width T & G	100 Lin. Ft. Covers Sq. Ft.	Ft. B. M. per 100 Sq. Ft.
2" x 8"	7¼"	62½	215	7¼"	60	220
3" x 8"	7¼"	62½	320	7¼"	60	330
2" x 10"	9½"	79	210	9¼"	67	215
3" x 10"	9½"	79	315	9¼"	77	320

Wood Bracing and Sheet Piling Banks By Hand

Quantities and Production are given per 100 sq. ft. of bank braced

Class of Work	Depth in Feet	Ft. B. M. Lbr. Req'd 100 Sq. Ft. Bank	No. Sq. Ft. Placed per 8-Hr. Day	Labor Hrs. Placing 100 Sq. Ft.	Labor Hrs. Removing 100 Sq. Ft.
Bracing Trenches	5- 8	100-150	150-200	4- 5	1 -1½
Trench Sheet Piling	5- 8	300-350	80- 90	9-10	2½-3
Trench Sheet Piling	10-15	325-375	65- 75	11-12	2½-3
Basement Sheet Piling	8-12	675-725	40- 50	17-18	5 -5½
Basement Sheet Piling	14-20	750-800	35- 45	20-21	5½-6

Some soil conditions, however, require pre- drilling or "hole-punching" before the wellpoints can be installed.

The discharge from the pumps should be piped to an area where it will not interfere with the construction work. The entire installation should be located so that it will interfere as little as possible with other divisions of the work.

It is sound practice to have a demonstrator from the wellpoint company on the job site to supervise the initial installation of the system. The cost of his time and expenses is more than compensated for by his working knowledge of wellpoint systems and their installation.

Operation of a wellpoint system is usually performed on a 24-hr. day, 7 day per week basis and must continue until all work, dependent on dry conditions, is completed.

02600   PAVING AND SURFACING

Concrete Curb and Gutter

When estimating the cost of forms for concrete curb or curb and gutter, the size and shape must be considered and the amount of materials estimated.

Combined curb and gutter is usually 12" high at the outside edge (including pavement thickness), 24" to 30" wide and 6" to 7½" thick.

Separate concrete curb is usually 6" wide at the top, 8" to 9" wide at the bottom, and 24", 30" or 36" thick.

Forms for Concrete Curb and Gutter.—When placing wood forms for separate concrete curb, two mechanics and a helper should set forms for 100 to 110 lin. ft. of 24" curb per 8-hr. day, at the following labor cost per 100 lin. ft.:

	Hours	Rate	Total	Rate	Total
Mechanics	15.0	$....	$....	$16.47	$247.05
Helper	7.5	....	....	12.54	94.05
Cost per 100 lin. ft.			$....		$341.10
Cost per lin. ft.			....		3.41

Two mechanics and a helper should set forms for 90 to 100 lin. ft. of 30" concrete curb per 8-hr. day, at the following labor cost per 100 lin. ft.:

	Hours	Rate	Total	Rate	Total
Mechanics	17.0	$....	$....	$16.47	$279.99
Helper	8.5	....	....	12.54	106.59
Cost per 100 lin. ft.			$....		$386.58
Cost per lin. ft.			....		3.87

When setting wood forms for combined concrete curb and gutter, two mechanics and a helper should set forms for 100 to

110 lin. ft. of curb and gutter per 8- hr. day, at the following labor cost per 100 lin. ft.:

	Hours	Rate	Total	Rate	Total
Cement mason	15.50	$....	$....	$16.47	$255.29
Helper..................	7.75	....	....	12.54	97.19
Cost per 100 lin. ft............................			$....		$352.48
Cost per lin. ft.			....		3.53

Wood forms for concrete curb and gutter are ordinarily built of 2" lumber. Including all bracing, it requires 2¼ to 2½ ft. of lumber, b.m. per sq. ft. of forms. This cost will run from 15 to 20 cents per sq. ft. and should be added to the above setting costs, taking into consideration the number of uses which are expected to be obtained from the forms. These forms may be used many times if properly taken care of.

Placing Ready-mix Concrete for Curb and Gutter.—A crew of 2 men should place about 1 cu. yd. of concrete an hr. at the following labor cost per cu. yd.:

	Hours	Rate	Total	Rate	Total
Labor	2	$....	$....	$12.54	$25.08

Concrete Roads and Pavements

Labor Placing Edge Forms for Concrete Paving.—Wood edge forms for concrete paving usually consist of 2" plank, of a width equal to the slab thickness, set to line and grade, secured to wood stakes and braced.

Two mechanics and a helper should set to line and grade from 250 to 300 lin. ft. of edge forms per 8-hr. day at the following labor cost per 100 lin. ft.:

	Hours	Rate	Total	Rate	Total
Mechanics..................	3.0	$....	$....	$16.47	$49.41
Helper..................	1.5	....	....	12.54	18.81
Cost per 100 lin. ft............................			$....		$68.22
Cost per lin. ft.			....		.68

Approximate Sq. Yds. of Concrete Pavement Obtainable from One Cu. Yd. of Concrete

Thickness Inches	Number Sq. Yds.	Thickness Inches	Number Sq. Yds.
4	9	8	4.5
4½	8	8½	4.25
5	7	9	4
5½	6.5	9½	3.75
6	6	10	3.5
6½	5.5	10½	3.5
7	5	11	3.25
7½	4.8	11½	3
		12	3

Placing Ready-mix Concrete Pavements.—Using ready-mix concrete and wheeling the concrete into place with wheelbarrows the labor per cu. yd. of concrete should cost as follows:

	Hours	Rate	Total	Rate	Total
Labor	1.2	$....	$....	$12.54	$15.05

If ready-mix concrete can be discharged from truck directly into place without wheeling, deduct 0.7-hr. labor time.

On larger jobs, containing 100 cu. yds. or more of concrete, better organization may be obtained and lower costs are the result. Assuming a pavement containing from 100 to 200 cu. yds. of ready-mixed concrete wheeled into place with buggies the labor per cu. yd. of concrete should cost as follows:

	Hours	Rate	Total	Rate	Total
Labor	.9	$....	$....	$12.54	$11.29

If ready-mix concrete can be discharged from truck directly into place without wheeling, deduct 0.4-hr. labor time.

Finishing Concrete Pavements.—Finishing concrete pavements usually consists of a series of operations performed in the following sequence: strike-off and consolidation, straightedging, floating, brooming and edging.

Labor Finishing Concrete Pavements.— Assuming an 8 inch concrete pavement being placed at the rate of 25 cu. yds. per hr. or 1,013 sq. ft. per hr. a crew of 10 cement masons should be able to keep up with the pour and perform all the above described finishing operations at the following labor cost per 100 sq. ft. :

	Hours	Rate	Total	Rate	Total
Cement Mason	1.0	$....	$....	$16.62	$16.62
Helper	.4	....	....	12.54	5.02
Cement mason foreman	.1	....	....	17.12	1.71
Cost per 100 sq. ft.			$....		$23.35
Cost per sq. ft.			....		.23

The above costs are based on a job which will run from 7,500 to 8,500 sq. ft. per 8-hr. day. On smaller jobs the costs will increase from 20 to 25 percent.

02800 LANDSCAPING

This item is usually sublet to a company specializing in this work, and items should be listed for the following:

Area of seeding and fertilizing.
Area of sodding.
Number, size and species of shrubs.
Number, size and species of trees.
Allowance for maintenance for specified period.
Allowance for guarantee, if specified.

For preliminary estimating purposes, however, the following information may be of some value.

The average price for seeding and fertilizing, including preparing the surface by disking or roto-tilling, grading and rolling, for small jobs, where much of the work is done by hand, ranges from 12 to 18 cts. per sq. ft. For the same operation on large jobs, where most of the work is done by machine, the average price is $1.35 per sq. yd.

Sodding may be done with either field sod or nursery sod and the operation includes preparing the surface as for seeding, fertilizing and rolling. For ordinary field sod the average price is from $1.60 to $1.80 per sq. yd. and for nursery sod the average price is from $2.50 to $2.75 per sq. yd.

CHAPTER 3

CONCRETE

CSI DIVISION 3

03100 CONCRETE FORMWORK

Concrete formwork should be estimated by computing the actual surface area of the form that comes into contact with the concrete; this is referred to as the "contact surface". The formwork quantities are then categorized into specific groups (wall form, column form, etc.) and in the proper unit of measure (square foot or linear foot) as follows:

Ribbon footing forms	-	Linear Foot
Slab edge forms (by height)	-	Linear Foot
Foundation edge forms	-	Square Foot
Wall Forms	-	Square Foot
Pier Forms	-	Square Foot
Column Forms	-	Square Foot
Beam bottom forms	-	Square Foot
Beam side forms	-	Square Foot
Supported slab forms	-	Square Foot

Forms for Concrete Footings.—There are two main types of footings which are used in most construction work, namely: column footings and wall footings. Column footings are usually square or rectangular masses of concrete designed to spread the column loading over an area to accommodate the soil bearing value and are located at the column centers. Wall footings are usually ribbons of concrete somewhat wider than the walls they carry and are similarly designed to spread the loading over a larger area than the actual wall bottom thickness.

Forms for footings which are formed completely, are usually simple and rough, using 2" planking for sides and 2"x4" stakes to hold the sides in place and 2"x4" struts for braces to the banks. This material may be reused numerous times in the course of the work.

Ribbon forms for partly formed footings usually consist of 2" thick lumber in as long lengths as practicable, secured at the correct elevation to 2"x4" stakes and tied across the tops at intervals of 4'-0" to 6'-0" with 1"x2" spreader-ties, except in the case of column footings which are usually diagonally tied at the corners with 1-inch boards and tied transversely, if required, with strap iron.

The following are itemized costs of the column and wall footing forms found in the average simple job. Complicated or "cut-up" jobs require up to 25 percent more material and up to 50 percent more labor.

Labor Cost of 100 Sq. Ft. of Column Footing Forms Based on
Footings 5'-0"x5'-0"x1'-6" Completely Formed

Placing Forms— 18-22 sq. ft. per hr.	Hours	Rate	Total	Rate	Total
Carpenter	5.0	$....	$....	$16.47	$ 82.35
Labor helping	2.5	....	....	12.54	31.35
Stripping Forms— 90-110 sq. ft. per hr.					
Labor	1.0	....	....	12.54	12.54
Cost per 100 sq. ft.			$....		$126.24
Labor cost per sq. ft.			....		1.26

Labor Cost of 100 Lin. Ft. of Column Footing Ribbon Forms Based on
Footings 5'-0"x5'-0"x1'-6" Partly Formed

Placing Forms— 35 Lin. Ft. per hr.	Hours	Rate	Total	Rate	Total
Carpenter	3.0	$....	$....	$16.47	$49.41
Labor helping	1.5	....	....	12.54	18.81
Stripping Forms— 450-500 Lin. Ft. per hr.					
Labor	0.21	....	....	12.54	2.64
Cost per 100 Lin. Ft.			$....		$70.86
Labor Cost per Lin. Ft.			....		.71

Labor Cost of 100 Sq. Ft. of Wall Footing Forms Based on
Footings 1'-0" Deep Completely Formed

Placing Forms— 25-30 sq. ft. per hr.	Hours	Rate	Total	Rate	Total
Carpenter	3.50	$....	$....	$16.47	$57.65
Labor	1.75	....	....	12.54	21.95
Stripping Forms— 70-75 sq. ft. per hr.					
Labor	1.50	....	....	12.54	18.81
Cost per 100 sq. ft.			$....		$98.41
Labor cost per sq. ft.			....		.98

Labor Cost of 100 Lin. Ft. of Wall Footing Ribbon Forms
Based on
Footings 1'-0" Deep Partly Formed

Placing Forms—
40

Lin. Ft. per hr.	Hours	Rate	Total	Rate	Total
Carpenter	2.5	$....	$....	$16.47	$41.18
Labor helping	1.25	....	....	12.54	15.68

Stripping Forms—
450-500

Lin. Ft. per Hr.					
Labor	0.21	....	....	12.54	2.64
Cost per 100 Lin. Ft.			$....		$59.50
Labor Cost per Lin. Ft.			....		.60

On jobs where the wall footings are stepped down to a lower elevation, a riser form is required at each step. The cost of this riser form is about twice the cost of the regular footing side form.

Grooves are usually formed in the tops of footings and are called "Footing Keys." They are formed by forcing a beveled 2"x4" into the surface of the wet concrete. After the concrete has hardened the "key" forms are pried loose, cleaned and reused.

Labor Cost of 100 Lin. Ft. 2"x4" Footing Keys

	Hours	Rate	Total	Rate	Total
Carpenter placing	1.25	$....	$....	$16.47	$20.59
Labor stripping	.50	....	....	12.54	6.27
Cost per 100 lin. ft.			$....		$26.86
Labor cost per lin. ft.			....		.27

Forms for Concrete Piers.—When column footing tops are lower than the bottom of the floor slab on the ground, piers are usually designed to carry the column loading to the footings.

Forms for this work are usually made up in side panels at the job bench, hauled to the footings, erected, clamped and braced.

Side panels are usually made of ⅝" plyform with 2"x4" stiffners. Clamping may be accomplished by using either 2"x4" lumber or metal column clamps properly spaced to withstand the pressure developed when placing the concrete.

Bracing is required to hold the form in position and usually consists of 2"x4" or 2"x6" lumber, secured to the form at grade level, spanning the pit and spiked to stakes driven into the ground. Bracing below this point is done by wedging 2"x4" struts between the form and the earth banks.

The following are itemized costs of making average pier forms; based on a pier 2'-0"x2'-0"x4'-0" containing 32 sq. ft.:

Labor Costs

Making Pier Forms	Hours	Rate	Total	Rate	Total
Carpenter	1.0	$....	$....	$16.47	$16.47
Labor	0.5	....	....	12.54	6.27
Cost 32 sq. ft. ...			$....		$22.74
Labor cost per sq. ft.				....	.71

Forms for Concrete Walls

Forms for concrete walls must be designed and constructed to withstand the horizontal pressures exerted by the fluid concrete against them. This pressure against any given point of the form will vary and be influenced by any one or all of the following factors:

1. Rate of vertical rise in filling the forms.
2. Temperature of the concrete and the weather.
3. Proportions of the mix and its consistency.
4. Method of placement and degree of vibration.

Concrete Wall Forms Made in Panels or Sections.—Many contractors, especially those performing smaller work, use foundation wall forms made up into panels 2'x6', 2'x8', 4'x6' or 4'x8' etc., in size. The frame for these sections is usually 2"x4" lumber, with intermediate braces spaced 12" to 16" on centers. The corners are reinforced with galvanized straps or strap iron, and the entire frame is sheathed with ⅝"or ¾" plyform plywood.

Forms of this kind can be used a large number of times, especially where the walls are fairly straight.

The labor cost of making panel forms will vary with the method used and the number made up at one time. A carpenter should complete one panel (28 to 32 sq. ft.) in ¾ to 1-hr. or 8 to 10 panels per 8-hr. day.

The cost of a panel 4'-0"x8'-0" in size, using ⅝" "Plyform" plywood should cost as follows:

Material Costs

	B.F.	Rate	Total	Rate	Total
2 pcs. 2"x4"-8'-0"	10.67	$....	$....	$ .36	$ 3.84
7 pcs. 2"x4"-4'-0"	18.67	....	....	.36	6.72
Plyform plywood, 4'-0"x8'-0"-⅝" sq. ft.	32			.50	16.00
Nails, etc.		....	....	.80	.80
Cost per panel 32 sq. ft.			$....		$27.36
Material cost per sq. ft.				....	.86

Labor Costs

	Hours	Rate	Total	Rate	Total
Carpenter	0.75	$....	$....	$16.47	$12.35
Labor	0.25	....	....	12.54	3.14
Cost per panel...			$....		$15.49
Labor cost per sq. ft.			....		.49

Built-In-Place Wood Forms For Concrete Walls

For those situations where the wall forms are to be built-in-place, the estimator should be familiar with form design parameters so that he can ascertain the correct quantity of form materials necessary for the work.

The following data and tables on form design have been furnished by The American Plywood Association of Tacoma, Washington:

Grade-Use Guide for Concrete Forms*

Use these terms when you specify plywood	DESCRIP-TION	VENEER GRADE	
		Faces	Inner Plies
APA B-B PLYFORM Class I & II**	Specifically manufactured for concrete forms. Many reuses. Smooth, solid surfaces. Mill-oiled unless otherwise specified.	B	C
APA High Density Overlaid PLYFORM Class I & II**	Hard, semiopaque resin-fiber overlay, heat-fused to panel faces. Smooth surface resists abrasion. Up to 200 reuses. Light oiling recommended between pours.	B	C Plugged

Grade-Use Guide for Concrete Forms*

Use these terms when you specify plywood	DESCRIP-TION	VENEER GRADE	
		Faces	Inner Plies
APA STRUCTURAL I PLYFORM**	Especially designed for engineered applications. All Group 1 species. Stronger and stiffer than PLYFORM Class I and II. Recommended for high pressures where face grain is parallel to supports. Also available with High Density Overlay faces.	B	C or C Plugged
Special Overlays, proprietary panels and Medium Density Overlaid plywood specifically designed for concrete forming**	Produce a smooth uniform concrete surface. Generally mill treated with form release agent. Check with manufacturer for design specifications, proper use, and surface treatment recommenda-tions for greatest number of reuses.		

*Commonly available in ⅝" and ¾" panel thicknesses (4'x8' size).

**Check dealer for availability in your area.

Concrete Pressures For Column And Wall Forms

| Pour Rate (ft./hr.) | Pressures of Vibrated Concrete (psf) (a), (b) | | | |
| | 50° F | | 70° F | |
	Columns	Walls	Columns	Walls
1	330	330	280	280
2	510	510	410	410
3	690	690	540	540
4	870	870	660	660
5	1050	1050	790	790
6	1230	1230	920	920
7	1410	1410	1050	1050
8	1590	1470	1180	1090
10	1950	1580	1440	1170

Notes: (a) Maximum pressure need not exceed 150h, where h is maximum height of pour.
(b) Based on concrete with density of 150 pcf and 4 in. slump.

Concrete Pressures For Slab Forms

| Depth of Slab (in.) | Concrete Pressure (psf) | |
	Non-Motorized Buggies (a)	Motorized Buggies (b)
4	100	125
5	113	138
6	125	150
7	138	163
8	150	175
9	163	188
10	175	200

Notes: (a) Includes 50 psf load for workmen, equipment, impact, etc.
(b) Includes 75 psf load for workmen, equipment, impact, etc.

Allowable Pressures On Plyform Class I For Architectural Applications
(deflection limited to 1/360th of the span)

Face Grain Across Supports Allowable Pressures (psf) (a)

Support Spacing (inches)	Plywood Thickness (inches)					
	1/2	5/8	3/4	7/8	1	1-1/8
4	2935	3675	4495	4690	5075	5645
8	970	1300	1650	1805	1950	2170
12	410	575	735	890	1185	1345
16	175	270	370	475	645	750
20	100	160	225	295	410	490
24			120	160	230	280
32					105	130
36						115

(a) Plywood continuous across two or more spans.

Face Grain Parallel To Supports Allowable Pressures (psf) (a)

Support Spacing (inches)	Plywood Thickness (inches)					
	1/2	5/8	3/4	7/8	1	1-1/8
4	1670	2110	2610	3100	4140	4900
8	605	810	1005	1190	1595	1885
12	215	360	620	740	985	1165
16		150	300	480	715	845
20		105	210	290	400	495
24			110	180	225	320

(a) Plywood continuous across two or more spans.

Estimating Material Requirements for Wood Forms Built-In-Place for Concrete Walls

The following examples of wall form material estimates assumes: 1. Concrete at 50° F with 4" slump, 2. Vertical pour rate of 3'-0" to 4'-0" per hour, and 3. A maximum form pressure of 750 PSF.

Lumber Required for Concrete Wall Forms 8'-0" High

Based on a wall 40'-0" long and 8'-0" high, containing 640 sq. ft. of forms.

	B.F.	Rate	Total	Rate	Total
Studs, 82 pcs. 2"x4"-8'-0" (12" on centers)	438	$....	$....	$.36	$157.68
Sheathing, ⅝" plyform sq. ft	640	....	....	.50	320.00
Plates, 80 lin. ft. 2"x4"	53			.36	19.08
Wales, 4 sets per side, 16 pcs. 2"x4"-40'-0".......	427			.36	153.72
Bracing and stakes, 2"x4"......	80		....	.36	28.80
Cost 640 sq. ft ...			$....		$679.28
Cost per sq. ft				....	1.06

Material not subject to reuse:	Quan.	Rate	Total	Rate	Total
Nails	30 lbs.	$....	$....	$.60	$18.00
Form ties (3,000 lb safe load) 24" o.c. wales.....	84		....	.40	33.60
Total cost 640 sq. ft....................			$....		$51.60
Cost per sq. ft			....		.08

Assuming 3 uses of form lumber, 1.06 ÷ 3=35.4¢ per use of form lumber, plus 8¢ per sq. ft. for nails and form ties, or a total of 43.4¢ per sq. ft. for material.

Lumber Required for Concrete Wall Forms 12'-0" High

Based on a wall 40'-0" long and 12'-0" high, containing 960 sq. ft. of forms.

	B.F.	Rate	Total	Rate	Total
Plates 4 pcs. 2"x4"-20'-0".......	53	$....	$....	$.36	$ 19.08
Studs, 82 pcs. 2"x4"-12'-0" (12" on centers)	656	....	....	.36	236.16
Sheathing, ⅝" plyform sq. ft	960	....	....	.50	480.00
Wales, 6 sets per side wales @ 24" on centers 24 pcs. 2"x 4"-40'-0".............	640	....	....	.36	230.40
Bracing and stakes, 2"x4"....	98	....	....	.36	35.28
Cost 960 sq. ft...			$....		$1,000.92
Cost per sq. ft			....		1.04

Maximum Spans for Lumber Framing, Inches*
Douglas Fir-Larch No. 2 Or Southern Pine No. 2 (KD)

Equivalent Uniform Load (lb/ft)	Continuous Over 2 or 3 Supports (1 or 2 Spans)								Continuous Over 4 or More Supports (3 or More Spans)							
	Nominal Size								Nominal Size							
	2x4	2x6	2x8	2x10	2x12	4x4	4x6	4x8	2x4	2x6	2x8	2x10	2x12	4x4	4x6	4x8
400	36	53	70	90	109	50	79	101	41	60	78	100	122	62	91	118
600	30	47	57	73	89	44	66	88	35	49	67	82	99	53	80	98
800	24	38	57	63	77	34	58	78	31	44	58	66	81	44	64	85
1000	21	33	43	55	67	35	51	68	29	39	51	57	69	39	58	76
1200	19	29	38	49	60	32	47	62	26	35	46	50	61	35	53	69
1400	17	27	35	45	54	30	43	57	24	32	42	46	56	31	49	64
1600	16	25	33	41	50	27	40	54	22	30	39	42	52	28	44	60
1800	15	23	30	39	47	25	38	51	21	28	36	40	48	26	40	55
2000	14	22	29	37	45	23	36	49	19	26	34	38	46	24	38	52
2200	13	21	28	35	43	22	34	46	18	25	33	36	44	23	36	50
2400	13	20	26	34	42	21	33	44	17	24	32	34	42	22	35	48
2600	12	19	25	33	40	20	31	42	16	23	31	33	40	21	33	46
2800	12	19	25	32	39	19	29	41	15	22	30	32	39	20	30	44
3000	12	18	24	31	38	18	28	38	14	20	29	31	38	19	29	38
3200	11	18	23	30	37	17	27	36	13	19	28	30	37	18	28	36
3400	11	17	23	30	36	17	26	35	13	18	27	30	36	17	27	35
3600	11	17	22	29	35	16	25	34	12	17	26	29	35	16	26	34
3800	11	17	22	28	35	16	24	33	12	17	25	29	35	16	25	33
4000	10	16	22	28	34	15	24	32	11	17	25	28	34	15	24	32
4500	10	16	21	27	32	15	22	30	11	16	24	28	32	14	23	30
5000	10	15	20	25	31	13	22	28	10	15	23	26	31	14	22	28

*Spans are based on PS-20 lumber sizes. Single member stresses were multiplied by a 1.25 duration-of-load factor for 7-day loads. Deflection limited to 1/360th of the psan with ¼" maximum. Spans are center-to-center of the supports.

Maximum Spans for Lumber Framing, Inches*
Hem-Fir No. 2

	Continuous Over 2 or 3 Supports (1 or 2 Spans) — Nominal Size								Continuous Over 4 or More Supports (3 or More Spans) — Nominal Size							
Equivalent Uniform Load (lb/ft)	2x4	2x6	2x8	2x10	2x12	4x4	4x6	4x8	2x4	2x6	2x8	2x10	2x12	4x4	4x6	4x8
400	33	48	63	80	97	48	73	96	36	53	70	90	109	56	81	107
600	27	39	51	65	79	41	59	78	27	43	57	68	89	45	66	88
800	22	34	44	56	69	35	51	68	22	35	46	59	72	39	58	76
1000	19	29	39	50	60	29	46	61	17	30	40	51	62	35	51	68
1200	16	26	35	44	54	27	42	55	16	27	36	45	55	31	47	62
1400	15	24	32	41	49	24	39	51	15	25	33	42	51	27	43	57
1600	14	23	30	38	46	22	36	48	15	23	30	39	47	25	39	52
1800	14	21	28	36	43	22	34	45	14	22	29	36	44	23	36	47
2000	13	20	27	34	41	21	31	43	13	21	27	35	42	21	33	44
2200	12	19	26	33	38	19	31	40	13	19	25	33	40	20	31	41
2400	12	18	25	31	37	18	29	38	12	18	25	32	39	19	30	39
2600	11	18	24	30	37	18	28	36	12	18	24	31	38	18	28	37
2800	11	18	23	30	36	17	26	35	11	17	24	30	37	17	27	36
3000	11	17	23	29	35	16	25	33	11	17	23	29	36	17	26	34
3200	10	17	22	28	34	16	24	32	11	17	22	29	35	16	25	33
3400	10	16	22	27	32	15	23	31	11	16	22	28	34	15	24	33
3600	10	16	21	27	32	15	22	30	10	16	21	27	33	15	23	31
3800	10	16	21	26	31	14	22	29	10	16	21	27	33	15	23	30
4000	10	15	20	25	31	14	21	29	10	16	20	27	32	14	22	29
4500	9	14	19	24	29	13	21	27	9	15	20	26	30	13	21	28
5000	9	13	18	23	28	12	20	26		15		25	30	13	20	26

*Spans are based on PS-20 lumber sizes. Single member stresses were multiplied by a 1.25 duration-of-load factor for 7-day loads. Deflection limited to 1/360th of the span with ¼" maximum. Spans are center-to-center of the supports.

Material not subject to reuse:	Quan.	Rate	Total	Rate	Total
Nails	40 lbs.	$....	$....	$.60	$24.00
Form ties (3,000 lb. safe load) 24" o.c. along wales	126	....	....	.40	50.40
Total cost 960 sq. ft.			$....		$74.40
Cost per sq. ft			....		.08

Assuming 3 uses of form lumber, 1.04 ÷ 3=34.7¢ per use for form lumber, plus 8¢ per sq. ft. for nail and form tiles, or a total of 42.7¢ per sq. ft. for material.

In comparing the preceeding example material estimates for wall forming on 8 and 12 foot high walls, you will note that both are in the same cost range. This will hold true for lower or higher walls as long as the design parameters for temperature, slump, pour rate, and allowable form pressure remain the same.

If the temperature of the concrete is lower, the mix is wetter, or it is advantageous to place the concrete into the forms at a faster pour rate, then the forms will have to be designed accordingly to withstand the increased pressure per square foot on the form surface.

The design changes would involve one or all of the following:

1. Stronger form tie (increased tensile strength).
2. Closer horizontal spacing for form ties.
3. Increased stud size to counteract bending under increased loads.
4. Closer on-center spacing of stud members to counteract deflection of the plyform, or, use thicker plyform material.
5. Increased size of wale members.

Labor Cost of 100 Sq. Ft. Built-in-Place Concrete Wall Forms 8'-0" High, Requiring 2.6 BF of Lumber Per Sq. Ft. of Forms

	Hours	Rate	Total	Rate	Total
Carpenter	4.50	$....	$....	$16.47	$ 74.12
Labor helping	2.30	....	....	12.54	28.84
Labor stripping-cleaning	2.20	....	....	12.54	27.59
Cost 100 sq. ft.			$....		$130.55
Labor cost per sq. ft.			....		1.31

Production Times for 100 Sq. Ft. of Wood Forms for Concrete Walls

Class of Work	Height Of Wall in Feet	BF. Lbr. per Sq. Ft. of Forms	No. Sq. Ft. of Forms 8-Hr. Day	Car-penter Hours	Labor* Hours	Removing Forms	
						No. Sq. Ft. 8-Hr. Day	Labor Hours
Wall Forms Built-in-Place	4 to 6	2.6	190-210	4.0	2.0	360	2.25
Wall Forms Built-in-Place	7 to 8	2.6	165-190	4.5	2.3	350	2.20
Wall Forms Built-in-Place	9 to 10	2.6	150-160	5.3	2.5	325	2.50
Wall Forms Built-in-Place	11 to 12	2.6	125-135	6.0	3.0	300	2.70

*If Union rules do not permit laborer working with carpenter, figure labor time as additional carpenter time.

Labor Cost of 100 Sq. Ft. Built-in-Place Concrete Wall
Forms 12'-0"
High, Requiring 2.6 BF of Lumber Per Sq. Ft. of Forms

	Hours	Rate	Total	Rate	Total
Carpenter.............	6.0	$....	$....	$16.47	$ 98.82
Labor helping........	3.0	....	....	12.54	37.62
Labor stripping-cleaning	2.7	....	....	12.54	33.86
Cost 100 sq. ft			$....		$170.30
Labor cost per sq. ft..........................			....		1.70

Pilasters.—Walls which are thickened for short intervals of 12" to 36" in length and then return to their former thickness are known as pilasters at that thickened area.

The cost of forming the face and sides of pilasters runs from 25 to 50 percent more than plain wall forms for both material and labor.

Radial Wall Forms.—Wall forms which are built to radii or other curves require closer stud spacing, more bracing and more labor laying out than straight wall forms.

Formwork for radial walls will cost from 50 to 100 percent more than forms for straight walls.

Forming Openings in Concrete Walls.—Openings in concrete walls for doors, windows, louvers, etc., are formed by placing bulkheads in the wall forms to form the sills, jambs and heads of the openings required. These forms usually require from 2¼ to 3 BF of lumber per sq. ft. of contact area, depending upon the size of the opening.

The following is an itemized labor cost of boxed opening forms, based on 4'-0"x4'-0" openings in a 1'-0" wall requiring 2½ BF of lumber per sq. ft. of contact area.

Labor Cost of 100 Sq. Ft. of Boxed Opening Forms

	Hours	Rate	Total	Rate	Total
Carpenter, setting forms..................	7.0	$....	$....	$16.47	$115.29
Labor Helping.......	3.5	....	....	12.54	43.89
Labor stripping forms..................	3.0	....	....	12.54	37.62
Cost 100 sq. ft...			$....		$196.80
Labor cost per sq. ft..........................			....		1.97

Forming Setbacks in Concrete Walls.—When concrete foundation walls extend some height above ground level, it is often required that the portion above grade be faced with masonry. This necessitates the portion above grade being reduced in thickness sufficient to accommodate the facing material.

This reduction in thickness is accomplished by placing a "setback' form in the regular wall forms.

Setback forms are usually made up of 2" framing lumber

faced with ⅝" or ¾" plyform, and usually require 1¾ to 2¼ BF of lumber per sq. ft. of contact area, depending mainly upon the thickness of the setback.

The following is an itemized labor cost of 100 sq. ft. of setback forms requiring 2 BF of lumber per sq. ft. of contact area:

	Hours	Rate	Total	Rate	Total
Carpenter	6.5	$....	$....	$16.47	$107.06
Labor helping	3.3	....	....	12.54	41.38
Labor stripping	2.7	....	....	12.54	33.86
Cost 100 sq. ft.			$....		$182.30
Labor cost per sq. ft.			....		1.82

Forming A Continuous Haunch on Concrete Walls—When an intermediate support on a wall face is required to provide bearing for masonry, slabs or future construction and reducing the wall thickness is prohibited by the design, a continuous haunch is commonly used.

A Continuous haunch is usually formed of 2" framing lumber, faced with ⅝" or ¾" plyform and requires from 3½ to 4 BF of lumber per sq. ft. of contact area.

The following is an itemized labor cost of 100 sq. ft. of continuous haunch forms:

	Hours	Rate	Total	Rate	Total
Carpenter	12.0	$....	$....	$16.47	$197.64
Labor helping	6.0	....	....	12.54	75.24
Labor stripping forms	2.5	....	....	12.54	31.35
Cost 100 sq. ft.			$....		$304.23
Labor cost per sq. ft.			....		3.04

Form Coating

For clean, easy stripping of forms and to prolong the life of form facings, the contact surface of all forms should receive a coating which will prevent concrete from bonding to the form face.

Forms should always be coated before erection, but where this is impracticable, such as for built-in-place wall forms, pan and joist forms for combination slabs, etc., oiling should be done in advance of setting reinforcing steel and care should be taken not to get any oil on concrete, masonry or steel bearing surfaces.

On large areas, such as flat slab decks, a laborer can coat 400 to 500 sq. ft. of surface per hour, but production is considerably less when coating forms for columns, beams, combination slabs, etc. A fair overall average for coating forms is about 300 to 350 sq. ft. per hour.

For better class work, when using plywood, plastic-coated plywood or lined forms, where smooth concrete surfaces are required, there are numerous patented coatings and lacquers available for this purpose. While the initial cost of these coatings is much higher than paraffin oil, better results are obtained and the forms may be used a couple of times without recoating, depending on the care used in applying the coating and in handling the forms when erecting or stripping.

Application may be either by brushing, spraying or dipping and a man can coat about 400 to 500 sq. ft. of form surface per hour.

Concrete Column Forms.

Reinforced concrete columns may be either square, rectangular, or round in shape. The forms for square or rectangular columns are usually constructed of wood while metal or fiber molds are used for round columns.

Forms for reinforced concrete columns should be estimated by the sq. ft. obtained by multiplying the girth of the column by the height. Example: obtain the number of sq. ft. of forms in a column 18" square and 12'-0" high. Proceed as follows: 18+18+18+18=72" or 6'-0", the girth. Multiply the girth by the height, 6'-0"x12'-0"=72 sq. ft. of forms.

The forms for all square or rectangular columns should be estimated in this manner.

To prevent the column forms from spreading while the concrete is placed and until it hardens, wood or metal clamps should be run around the column at intervals, depending upon the height of the column.

In recent years the use of wood clamps has been almost entirely eliminated by the use of metal clamps.

The pressure of the fresh concrete has a greater effect on the design of column forms than on wall forms on account of the smaller amount of concrete to be placed. The contractor usually finds it more convenient and economical to fill a group of medium size columns in as short time as possible, perhaps 30 minutes, when the full liquid pressure of the fresh concrete will be acting: namely, 150 lbs. per sq. ft. for every foot in depth of the concrete.

Labor Cost of 100 Sq. Ft. Concrete Column Forms Using Plywood Sheathing with Adjustable Steel Clamps and 2"x4" Vertical Cleats

Making Forms	Hours	Rate	Total	Rate	Total
Carpenter	4.0	$....	$....	$16.47	$ 65.88
Labor	2.0	....	....	12.54	25.08
Assembling, Erecting, Plumbing and Bracing					
Carpenter	3.0	....	....	16.47	49.41
Labor helping	3.0	....	....	12.54	37.62
Removing Clamps and Forms					

Labor	2.0	....	....	12.54	25.08
Cost per 100 sq. ft			$....		$203.07
Labor cost per sq. ft				....	2.03

Sonotube* Fibre Forms.—Sonotube fibre forms are spirally wound, laminated fibre tubes which were developed to provide a fast and economical method of forming for round concrete columns, piers, etc., and particularly adaptable as forms for encasing wood or steel piling. They provide a fast, economical method of forming all types of round concrete columns. They also permit new design possibilities for architects, provide new structural features for engineers, and save time, labor and money.

Sonotube fibre forms are available in standard sizes from 8" to 48" inside diameter, with wall thickness ranging from .200" to .500"

They may be obtained in any length up to 48 ft. and may be easily cut to length on the job with an ordinary hand or power saw.

There are three types of Sonotube fibre forms to meet virtually any combination of job requirements. The seamless is the premium form with specially finished inner ply which produces a smoother, continuous concrete surface. Regular A-coated is the standard form for exposed columns. For exposed columns not exceeding 12 ft. in height, the light wall A-coated is recommended.

Sonotube fibre forms are lightweight, easily handled and require few men for erection manually or with simple block and tackle. Minimum bracing is required, only fixing in place at bottom and light lumber bracing to keep them erect and plumb. No clamping is necessary.

Under normal conditions, for average size columns, 12" to 18" diameter, 10-ft. to 14-ft. in length, a carpenter and a helper should handle, cut to length, erect and brace about 12 Sonotube column forms per 8-hr. day at the following labor cost per column:

	Hours	Rate	Total	Rate	Total
Carpenter	0.7	$....	$....	$16.47	$11.53
Helper	0.7	....	....	12.54	8.78
Cost per column			$....		$20.31

Sonotube fibre forms strip easiest and most economically from two to ten days after concrete is placed. Stripping may be accomplished with the least effort by using an electric hand saw, with blade set to cut slightly less than the wall thickness of Sonotube, and make two or three evenly spaced vertical cuts from bottom to top, after which form may be pulled free from column. Sonotube forms are for one time use only.

Under normal conditions, a carpenter and a helper should

Roos Adjustable Metal Column Clamps
Hinged Bar Type—2 Hinged Units=1 Complete Clamp

Clamp Size	Bar Size	Weight Per Clamp	Net Concrete Sizes			
			With 1" Column Lumber		With 2" Column Lumber	
			Minimum	Maximum	Minimum	Maximum
36"	5/16" x 21/2" x 36"	40 lbs.	81/2" x 81/2"	26" x 26"	61/2" x 61/2"	24" x 24"
48"	3/8" x 21/2" x 48"	58 lbs.	12" x 12"	38" x 38"	10" x 10"	36" x 36"
60"	3/8" x 3" x 60"	85 lbs.	221/2" x 221/2"	481/2" x 481/2"	201/2" x 201/2"	461/2" x 461/2"
36" x 60"	1 Leg, 5/16" x 21/2 x 36" 1 Leg, 3/8" x 3" x 60"	65 lbs.	71/2" (36" Leg) 231/2" (Leg)	25" (36" Leg) 491/2" (60" Leg)	51/2" (36" Leg) 211/2" (60" Leg)	23" (36" Leg) 471/2" (60" Leg)

strip 25 to 30 average size Sonotube column forms per 8-hr. day at the following labor cost per column:

	Hours	Rate	Total	Rate	Total
Carpenter	0.3	$....	$....	$16.47	$4.94
Helper	0.3	....	....	12.54	3.76
Cost per column			$....		$8.70

FORMS FOR CONCRETE BEAMS, GIRDERS, SPANDREL BEAMS, AND LINTELS

Forms for reinforced concrete beams, girders, lintels, etc., should be estimated by the sq. ft. obtained by adding the dimensions of the three sides of the beam and multiplying by the length. Example: a beam or girder may be marked 12"x24" on the plans but perhaps 6" of this depth is included in the slab thickness, so the form area should be obtained by taking the dimensions of the beam or girder from the underside of the slab, i.e., 18"+12"+18"=48" or 4'-0" the girth, which multiplied by the length gives the number of sq. ft. of forms in the beam, viz., 4'-0"x18'-0"=72 sq.ft.

Spandrel beams or those projecting above or below the slab, should be measured in the same manner. A good rule to remember when measuring beam forms is to take the area of all forms that come in contact with the concrete.

Labor Cost of Wood Forms for Concrete Beams and Girders.

The labor cost on beam and girder forms will vary considerably on different jobs, due to the amount of duplication, size of beams and girders and the methods of framing and placing them.

It requires 4 to 4½ B.F. of lumber to complete 1 sq. ft. of beam and girder forms, which includes uprights or "shores", beam soffits, bracing, etc.

A carpenter should frame and erect 250 to 275 B.F. of lumber per 8-hr. day, at the following labor cost per 1,000 BF:

	Hours	Rate	Total	Rate	Total
Carpenter	29.0	$....	$....	$16.47	$477.63
Helper	14.5	....	....	12.54	181.83
Cost per 1,000 BF			$....		$659.46

If laborers are not permitted to carry the lumber and forms for the carpenters, the "helper" time given above should be figured as carpenter time.

A laborer should remove or "strip" 900 to 1,000 BF of lumber per 8-hr. day, at the following labor cost per 1,000 BF:

	Hours	Rate	Total	Rate	Total
Labor	9	$....	$....	$12.54	$112.86

Forms for Upturned Concrete Beams.—When the top of a beam is higher than the floor slab level, the beam side forms for that portion above the floor must be supported on temporary legs which are removed after the concrete is poured and is still wet. It is also necessary to use spreaders to maintain the proper beam width.

While it is true that the additional form material required is negligible, the labor cost on this kind of beam side form will run 50 to 100 percent more than ordinary beam side forms.

FORMS FOR REINFORCED CONCRETE FLOORS

Forms for all types of reinforced concrete floors should be estimated by the sq. ft. taking the actual floor area or the area of the wood or metal forms that come in contact with the concrete.

Designing Forms for Reinforced Concrete Floors and Estimating Quantity of Lumber Required

The process of designing forms for reinforced concrete floors may be divided into several parts, such as determining the span of the sheathing; size and spacing of joists and girders or stringers, and determining the size and spacing of the upright supports or shores.

The spacing of joists for concrete floor forms is governed by the strength of the plywood used.

When plywood panels are used for sheathing, the spacing of the joists must coincide with the sizes of the plywood panels, 4'-0"x8'-0", which means that the joists should be spaced 12", 16" or 19" on centers, depending upon the load to be carried.

Weight of Concrete Floors of Various Thicknesses

In determining the weight per sq. ft. of floor, wet concrete is figured at 150 lbs. per cu. ft. Dead load of form lumber and live load on forms while concrete is being placed is figured at 38 lbs. per sq. ft. which is sufficient for temporary construction.

Floor Thickness..	2"	3"	4"	5"	6"	7"	8"	9"	10"	11"	12"
Conc. Wt. per Sq. ft., Lbs	25	37.5	50	62.5	75	87.5	100	112.5	125	137.5	150
Temp. Live Load per Sq. ft., Lbs	40	37.5	38	37.5	38	37.5	38	37.5	38	37.5	38
Total Weight per Sq. Ft., Lbs	65	75.0	88	100.0	113	125.0	138	150.0	163	175.0	188

Data on Plywood Sheathing

When plywood is used for sheathing, the following loads may be safely carried.

ALLOWABLE PRESSURES ON PLYFORM
CLASS I FOR ARCHITECTURAL APPLICATIONS
(deflection limited to 1/360th of the span)

FACE GRAIN ACROSS SUPPORTS
ALLOWABLE PRESSURES (psf) (a)

Support Spacing (inches)	PLYWOOD THICKNESS (inches)					
	½	⅝	¾	⅞	1	1-⅛
4	2935	3675	4495	4690	5075	5645
8	970	1300	1650	1805	1950	2170
12	410	575	735	890	1185	1345
16	175	270	370	475	645	750
20	100	160	225	295	410	490
24			120	160	230	280
32					105	130
36						115

(a) Plywood continuous across two or more spans.

Carrying Capacity of 4"x4" Uprights or Shores.—The carrying capacity of a given size upright or shore is generally limited by two principal factors. The first is compression at right angles to the fibers which the shore exerts on the cross beam or stringer. The stress per sq. in. on the bearing area of the stringer on the shore should not greatly exceed 500 lbs. per sq. in. for Yellow Pine, Western Fir or similar lumber, otherwise there is a noticeable impression of the shore into the fibers of the stringer, especially when the lumber is green or water soaked after a long rain. This often amounts to ⅛". Hence a (4"x4", S4S), 3½"x3½" upright should never be loaded to more than 6,000 lbs., no matter how short it is, and a 4"x4" rough shore should not be loaded to more than 8,000 lbs.

The other principal limitation of strength is due to the length of the shore and the degree of crookedness in its length. Every experienced contractor knows that shores often show bows of 1" to 2" and if it is further considered that the shores are not cut off exactly square to their length, the carrying capacity of most shores are greatly limited by the eccentric loading impressed on them by the uneven bearing of the stringer and the bow in the shores.

The following table gives the permissible load on shores for various lengths and eccentricities:

Safe Load in Pounds on Uprights or Shores of Various Lengths

Length of Upright or Shore Eccentricity	4"x4" S4S Yellow Pine or Fir 8'-0"	10'-0"	4"x4" Rough Yellow Pine or Fir 8'-0"	10'-0"	4"x4" Rough Wisconsin Hemlock 8'-0"	10'-0"
(Bow+Top) 0"	6,000	6,000	8,000	8,000	8,000	8,000
(Bow+Top) 1"	5,700	4,800	8,000	7,100	7,300	5,600
(Bow+Top) 1½"	4,700	3,900	7,300	5,800	6,000	5,100
(Bow+Top) 2"	4,200	3,400	6,700	5,500	5,200	4,400
(Bow+Top) 2½"	3,600	3,000	5,300	4,500	4,800	4,100
(Bow+Top) 3"	3,400	2,800	5,100	4,300	4,500	3,800

Adjustable Shores for Reinforced Concrete Floors

There are a number of adjustable shores on the market to support reinforced concrete beam and slab forms. These shores are a combination of wood and metal, or all metal, which can be raised or lowered within certain limits.

Adjustable Shores Versus 4"x4's.—It is much easier to shore up a floor using adjustable shores instead of 4"x4"s, which require wedging and cutting or adding to for each change in story height. This additional labor all costs money at present wage scales. However, to offset this, the original cost of the adjustable shores is much more than the 4"x4"s but if they can be used often enough they will eventually show a saving. This will have to be considered when contemplating a purchase or rental basis.

Another thing that should be considered is the fact that a 4"x4" shore will usually carry a larger load than an adjustable shore, so in many cases it will require more adjustable shores than 4"x4"s to carry the same load. On the other hand, especially on light constructed floors, the adjustable shores may carry all the load required. This is an item that should be considered by the estimator before pricing the job.

Labor cost of Flat Slab Forms.—Forms for flat slab floors are continuous— without breaks or offsets—except at column heads, stair wells, elevator shafts, etc., where it is necessary to frame for beams and girders.

The sq. ft. labor cost will vary with the amount of lumber required per sq. ft. of forms, as the heavier the slab and the higher the ceiling, the more lumber required per sq. ft.

Forms for exceedingly light floor slabs cost proportionately more per sq. ft. than heavy forms as the preliminary work is the same with both types. However, on work of this class a carpenter should frame and erect 375 to 425 BF of lumber per 8-hr. day, at the following labor cost per 1,000 BF:

	Hours	Rate	Total	Rate	Total
Carpenter	20	$....	$....	$16.47	$329.40
Labor	10	....	....	12.54	125.40
Cost per 1,000 BF			....	$....	$454.80

When removing or "stripping" forms, a laborer should remove 1,000 to 1,200 BF of lumber per 8-hr. day, at the following labor cost per 1,000 BF:

	Hours	Rate	Total	Rate	Total
Labor	7	$....	$....	$12.54	$87.78

Forms for Beam and Girder Type Solid Concrete Floors

The labor cost of wood forms for beam and girder type floors will run somewhat higher than flat slab floors on account of the shorter spans, additional framing around beams, girders, etc.

On forms of this class, a carpenter should frame and erect 325 to 375 BF of lumber per 8-hr. day, at the following labor cost per 1,000 BF:

	Hours	Rate	Total	Rate	Total
Carpenter	23.0	$....	$....	$16.47	$378.81
Labor	11.5	....	....	12.54	144.21
Cost per 1,000 BF			$....		$523.02

An experienced laborer should remove or "strip" 1,000 to 1,200 BF. of lumber, per 8-hr. day, at the following labor cost per 1,000 BF.:

	Hours	Rate	Total	Rate	Total
Labor	7	$....	$....	$12.54	$87.78

Forms for Floors of Metal Pan and Concrete Joist Construction

Wood forms for floors of metal pan and concrete joist construction may be of much lighter construction than solid concrete floors, as the dead load is less.

Metal pans are usually furnished 20" to 30" wide, with 2"x6" or 2"x8" plank spaced 24" to 37" on centers, depending upon the width of the joists, which vary from 4" to 7".

The open deck construction is ordinarily used owing to the wide spacing of the floor boards.

In buildings having one-way beams and long floor spans, a carpenter should frame and erect 350 to 400 BF of lumber per 8-hr. day, at the following labor cost per 1,000 BF:

	Hours	Rate	Total	Rate	Total
Carpenter	21.4	$....	$....	$16.47	$352.46
Labor	10.7	....	....	12.54	134.18
Cost per 1,000 BF			$....		$486.64

Floors having beams and girders running in both directions with the floor panels averaging 16'-0"x16'-0" and smaller, a carpenter should frame and erect 325 to 375 BF of lumber per 8-hr. day, at the following labor cost per 1,000 BF:

	Hours	Rate	Total	Rate	Total
Carpenter	22.8	$....	$....	$16.47	$375.52
Labor	11.4	....	....	12.54	142.96
Cost per 1,000 BF			$....		$518.48

An experienced laborer should remove or strip 1,000 to 1,200 BF. of lumber per 8-hr. day, at the following labor cost per 1,000 BF:

	Hours	Rate	Total	Rate	Total
Labor	7	$....	$....	$16.47	$87.78

Number of Feet of Lumber Required for One Sq. Ft. of Floor Forms for Metal Pan and Concrete Joist Construction

The following table is based on the assumption the ends of joists and stringers are supported by the beam and girder forms.

Ceiling Height in Feet	Thickness of Floor Slab in Inches				
	8½" 85 lbs.	10½" 90 lbs.	12½" 100 lbs.	14½" 105 lbs.	16½" 115 lbs.
10	1.7	1.7	1.7	1.8	1.8
12	1.7	1.7	1.8	1.9	1.9
14	1.7	1.7	1.8	1.9	2.0
16	1.8	1.8	1.9	2.0	2.1
18	2.1	2.1	2.2	2.3	2.4
20	2.2	2.2	2.3	2.4	2.5

Number of Feet of Lumber Required for One Sq. Ft. of Floor Forms For Metal Pan and Concrete Joist Construction

The following table is based on the assumption it is a wall bearing job and will require wood girts or stringers together with supporting shores at both ends of joists and stringers.

Ceiling Height in Feet	Thickness of Floor Slab in Inches				
	8½" 85 lbs.	10½" 90 lbs.	12½" 100 lbs.	14½" 105 lbs.	16½" 115 lbs.
10	2.1	2.1	2.1	2.2	2.2
12	2.1	2.1	2.1	2.2	2.2
14	2.2	2.2	2.2	2.3	2.4
16	2.3	2.3	2.3	2.4	2.4
18	2.5	2.5	2.6	2.7	2.8
20	2.6	2.6	2.7	2.8	2.9

FORMS FOR REINFORCED CONCRETE STAIRS

Wood Forms for Straight Concrete Stairs.—On straight stairs extending from floor to floor without intermediate landing platforms and having an average story height of 10'-0" to 12'-0", two carpenters working together should lay out the stair, place rough stringers, mark off treads and risers on the rough string boards and set them in place in about 8 to 9 hours.

After the rough strings are in place it will require another 8 to 10 hrs. for 2 carpenters to sheath the stairs, cut, bevel and place risers, and place all necessary shoring and bracing ready for concrete.

The forms for an average flight of concrete stairs, 4'-0" wide and 18'-0" long, containing 72 sq. ft. of forms (soffit measurement) or 72 lin. ft. of risers, should cost as follows:

	Hours	Rate	Total	Rate	Total
Carpenter	36	$....	$....	$16.47	$592.92
Labor helping and removing forms	9	....	....	12.54	112.86
Cost per flight			$....		$705.78
Cost per sq. ft.				....	9.80
Cost per lin. ft. riser				....	9.80
Cost per riser (18)				....	39.21

The lumber cost for this stair should run about as follows:

	BF	Rate	Total
Stringers, 2 pcs. 2"x12"-20'-0"	80	$0.36	$ 28.80
Risers, 18 pcs. 2"x8"-4'-0"	96	.36	34.56
Soffit sheathing, 4'-0"x18'-0"	72	.50	36.00
Joists supporting sheathing, 4 pcs. 2"x8"-18'-0"	96	.36	34.56
Shores or uprights, 6 pcs. 4"x4"-10'-0"	80	.36	28.80
Purlins or stringers, 3 pcs. 4"x6"-4'-0"	24	.36	8.64
Sills, wedges and bracing	30	.33	9.90
Cost per flight			$181.26
Cost per sq. ft. (72)			2.52

FORMS OTHER THAN WOOD

Lightweight Steel Forming Material for Concrete.—High-strength corrugated steel forming material is often used for forming reinforced concrete floor and roof slabs. It comes either galvanized or uncoated and is manufactured from 100,000 p.s.i. tough-temper steel. It has a definite reliable structural strength nearly twice that of ordinary steel having equal weight. Used primarily in floor and roof systems having steel joists, junior beams, or purlins it is also used over pipe tunnels or similar installations where economies of using permanent forms can be realized. Either structural grade concrete or lightweight, insulating concrete may be used with corrugated steel forms, but in either case reinforcing bars or wire mesh should be added to satisfy flexure and temperature steel requirements.

Erection Costs.—Corrugated steel forming material can be rapidly erected in all kinds of weather thereby providing cover to floors below. The corrugated sheets are large in area but extremely light in weight and can be easily handled by one man.

Using an average figure of 4,000 square feet per eight hour day for a four man crew, the labor costs for placing 1,000 square feet of corrugated steel forming material would be as follows:

	Hours	Rate	Rate	Total
Ironworker	8	$.... $....	$17.70	$141.60
Cost per sq. ft		....		.14

The above figures are based upon one man welding or fastening the deck in place and three men handling and placing the material.

Computing Concrete Quantities.—When computing volume of concrete placed over corrugated steel decking, subtract ¼-in. from slab thickness to allow for corrugations of standard and ⅜-in. for heavy duty corrugated steel forming. Size of corrugations; standard, ½-in. depth x 23/16-in. pitch; heavy duty, ¾-in. depth x 3-in. pitch.

03200   CONCRETE REINFORCEMENT

Reinforcing steel is estimated by the pound or the ton, obtained by listing all bars of different sizes and lengths and extending the total to pounds. Reinforcing bars may be purchased from warehouse in stock lengths and all cutting and bending done on the job or they may be purchased cut to length and bent, ready to place in the building.

Standard Sizes and Weights of Concrete Reinforcing Bars

Bar Designation (Numbers)	Unit Weight Pounds per Foot	Nominal Dimensions,—Round Sections Inches Diameter	Cross Sectional area Sq. In.	Perimeter Inches
2	0.167	.250	0.05	0.786
3	0.376	.375	0.11	1.178
4	0.668	.500	0.20	1.571
5	1.043	.625	0.31	1.963
6	1.502	.750	0.44	2.356
7	2.044	.875	0.60	2.749
8	2.670	1.000	0.79	3.142

9	3.400	1.128	1.00	3.544
10	4.303	1.270	1.27	3.990
11	5.313	1.410	1.56	4.430
14S	7.650	1.693	2.25	5.320
18S	13.600	2.257	4.00	7.090

Bar No. 2 in plain rounds only.

Labor Placing Reinforcing Bars.—The labor cost of placing reinforcing steel will vary with the average weight of the bars, the manner in which it is placed; i.e., whether bars may be laid loose or whether it is necessary to tie them in place, and upon the class of labor employed.

Setting Reinforcing Bars.—On jobs using bars ⅝" and smaller, where it is not necessary to tie them in place, experienced reinforcing steel setters should handle and place 900 to 1,100 lbs. of steel per 8-hr. day, at the following labor cost per ton:

	Hours	Rate	Total	Rate	Total
Iron worker	16	$....	$....	$17.58	$281.28

On jobs using bars ¾" and heavier, where it is not necessary to tie the bars in place, an experienced reinforcing steel setter should handle and place 1,400 to 1,600 lbs. of steel per 8-hr. day, at the following labor cost per ton:

	Hrs	Rate	Total	Rate	Total
Iron worker	10.6	$....	$....	$17.58	$186.35

Setting Reinforcing Bars Tied in Place.—On jobs having concrete floors where it is necessary to tie the bars in place with tie wire, and where lightweight bars, (⅝" and under) are used, an experienced reinforcing steel setter should handle, place and tie 800 to 1,000 lbs. of steel per 8-hr. day, at the following labor cost per ton:

	Hours	Rate	Total	Rate	Total
Iron worker	17.8	$....	$....	$17.58	$312.93

On jobs using heavy bars, (¾" and over) an experienced reinforcing steel setter should handle, set and tie 1,300 to 1,500 lbs. per 8-hr. day, at the following labor cost per ton:

	Hours	Rate	Total	Rate	Total
Iron worker	11.5	$....	$....	$17.58	$202.17

Welded Steel Fabric or Mesh Reinforcing

Welded steel fabric is a popular and economical reinforcing for concrete work of all kinds, especially driveways and floors. It may also be used for temperature reinforcing, beam and column wrapping, road and pavement reinforcing, etc.

It is usually furnished in square or rectangular mesh.

It is usually sold at a certain price per sq. ft. or sq. yd. depending upon the weight.

Prices of mesh vary with quantity ordered and freight from mill to destination. Typical prices for some styles usually stocked, with truck job site delivery are as follows: Approximate prices are per 100 sq. ft.

Style	Wt. Per 100 Sq. Ft.	Under 5 Tons
6x6-6x6 (W2.9)	42 lbs.	$15.00
6x6-8x8 (W2.1)	30 lbs.	10.50
4x4-8x8 (W2.1)	44 lbs.	15.75

Labor Placing Mesh Reinforcing In Walls.—When mesh reinforcing is used, furnished in sheets or rolls, a man should place 700 to 800 sq. ft. in walls per 8-hr. day, at the following labor cost per 100 sq. ft. :

	Hours	Rate	Total	Rate	Total
Steel setter	1.0	$....	$....	$17.58	$17.58
Cost per sq. ft			....		.176

Labor Placing Steel Fabric on Floors.—When used for reinforcing floor slabs or for temperature reinforcing, a man should place 1,400 to 1,600 sq. ft. of fabric per 8-hr. day, at the following labor cost per 100 sq. ft. :

	Hours	Rate	Total	Rate	Total
Iron worker	0.53	$....	$....	$17.58	$9.32
Cost per sq. ft			....		.093

03300   CAST-IN-PLACE CONCRETE

Weights of Miscellaneous Building Materials
Weight of Cement, Sand, Gravel, Crushed Stone, Concrete, Etc.

The following table gives the approximate weights of cement, concrete and concrete aggregates, and will assist the estimator in figuring freight or when purchasing materials by weight:

	Weight Lbs. Per Cu. Yd.
Bank sand	2,500
Torpedo sand	2,700
Crushed stone	2,500
Crushed stone screenings	2,500
Gravel	2,700
Roofing gravel	2,700
Cement weighs 376 lbs. per bbl.	
(1-sack weighs 94 lbs. and bulks 1-cu. ft.)	
Cinder concrete.............................Wt. per cu. ft.	112
Concrete of gravel, limestone, sandstone,	
trap rock, etc. ....Wt. per cu. ft.	150

CONCRETE

Concrete should be estimated by the cu. ft. or cu. yd. containing 27 cu. ft. The cu. yd. is the most convenient unit to use because practically all published tables giving quantities of cement, sand, gravel or crushed stone are based on the cu. yd. of concrete. The cu. yd. is also a convenient unit to use when estimating labor costs.

The Importance of Mixing Water.—The hardening of concrete mixtures is brought about by chemical reactions between cement and water, the aggregates (sand, gravel, stone, etc.) being inactive ingredients used as fillers. So long as workable mixtures are used, the less water there is in the mix, the stronger, more watertight and more durable will be the concrete. Excess mixing water dilutes the paste made by the cement and water and makes weaker, more porous concrete. For any set of conditions of mixing, placing and curing and for given materials there is a definite relation between the strength of the concrete and the amount of water used in mixing. This relation can be determined by test and a water-cement ratio- strength curve can be developed representing the actual results on a specific job.

Table 1—Recommended Slumps
For Various Types of Construction

Types of construction	Slump, in.	
	Maximum*	Minimum
Reinforced foundation walls and footings	3	1
Plain footings, caissons, and substructure walls	3	1
Beams and reinforced walls	4	1
Building columns	4	1
Pavements and slabs	3	1
Heavy mass concrete	3	1

*May be increased 1 in. for methods of consolidation other than vibration.

READY-MIXED CONCRETE

In the United States more than 90 percent of all concrete being used in building projects is ready-mixed. Ready-mixed concrete may be defined as portland cement concrete manufactured for delivery to a purchaser in a plastic and unhardened state.

Truck mixers consist essentially of a mixer with a separate water tank and water measuring device mounted on a truck chassis. There are other truck conveyances which are similar but are without provisions for water.

ASTM C-94 requires that when a truck mixer is used either for complete mixing or to finish the partial mixing, each batch of concrete is to be mixed not more than 100 revolutions of the drum or blades at the speed of rotation designated by the manufacturer as the mixing speed. Any additional mixing is to be done at the agitating speed. The specification also requires that the concrete must be delivered and discharged from the truck mixer or agitator truck within $1\frac{1}{2}$ hours after introduction of the water to the cement and aggregate or the cement to the aggregate.

High Early Strength Portland Cement.—Most of the cement companies are now manufacturing what is designated as high early strength portland cement, which develops practically the same strength in 72 hours as is obtained with normal portland cement in 7 to 10 days.

Air Entrained Concrete.—Most of the highway departments are specifying air-entrained concrete for pavements and many require it for bridges and other structures. In this concrete 3 to 5 or 6 percent of air is incorporated in the concrete in the form of minute separated air bubbles. Such concrete is more resistant to freezing and thawing and to salt action than normal concrete. It is produced by using air-entraining portland cement or by the addition of an air-entraining agent at the mixer. Originally developed to prevent scaling of pavement where salts are used for ice removal, air-entrained concrete is being widely adopted for all types of work and in all locations because of its better workability as well as its better resistance to weathering even where salts are not used.

Mixes for Small Jobs

For small jobs where time and personnel are not available to determine proportions in accordance with the recommended procedure, mixes in Table 1 will usually provide concrete that is amply strong and durable if the amount of water added at the mixer is never large enough to make the concrete overwet. These mixes have been predetermined in conformity with the recommended procedure by assuming conditions applicable to the average small job, and for aggregate of medium specific gravity. Three mixes are given for each maximum size of coarse aggregate. For the selected size of coarse aggregate, Mix B is intended for initial use. If this mix proves to be oversanded, change to Mix C; if it is undersanded, change to Mix A. It should be noted that the mixes listed in the table are based on dry or surface-dry sand. If the sand is moist or wet, make the corrections in batch weight prescribed in the footnote.

The approximate cement content per cubic foot of concrete listed in the table will be helpful in estimating cement requirements for the job. These requirements are based on concrete that has just enough water in it to permit ready working into forms without objectionable segregation. Concrete should slide, not run, off a shovel.

Table 1—Concrete Mixes for Small Jobs

Maximum size of aggregate, in.	Mix designation	Approximate weights of solid ingredients per cu. ft. of concrete, lb.				
			Sand.*		Coarse aggregate	
		Cement	Air-entrained concrete†	Concrete without air	Gravel or crushed stone	Iron blast furnace slag
½	A	25	48	51	54	47
	B	25	46	49	56	49
	C	25	44	47	58	51
¾	A	23	45	49	62	54
	B	23	43	47	64	56
	C	23	41	45	66	58
1	A	22	41	45	70	61
	B	22	39	43	72	63
	C	22	37	41	74	65
1½	A	20	41	45	74	65
	B	20	39	43	77	67
	C	20	37	41	79	69
2	A	19	40	45	79	69
	B	19	38	43	81	71
	C	19	36	41	83	72

*Weights are for dry sand, If damp sand is used, increase tabulated weight of sand 2 lb. and, if very wet sand is used, 4 lb.

†Air-entrained concrete should be used in all structures which will be exposed to alternate cycles of freezing and thawing. Air-entrainment can be obtained by the use of an air-entraining cement or by adding an air-entraining admixture. If an admixture is used, the amount recommended by the manufacturer will, in most cases, produce the desired air content.

Capacities of Concrete Mixers Based on A. G. C. Standards
(in Sacks of Cement per Batch)

Size of mixer denotes number of cubic feet of mixed concrete, plus 10 percent excess, when mixer is level.

Concrete Proportions	Sizes of Standard Mixers				
	3½ - S	6 - S	11 - S	16 - S	28 - S
1:1½:3	½	1	3	4	8
1:1½:3½	½	1	2	3	7
1:2 :3	½	1	2	3	7
1:2 :3½	½	1	2	3	7

Concrete Proportions	Sizes of Standard Mixers				
	3½ - S	6 - S	11 - S	16 - S	28 - S
1:2 :4	½	1	2	3	6
1:2½:4	½	1	2	3	6
1:3 :5	½	1	1	2	5
1:3 :6	½	1	1	2	4

Labor Mixing and Placing One Cu. Yd. of Concrete for Foundations

Method of Mixing and Placing	Hours per Cubic Yard			
	Common Labor	Mixer Engr.	Hoist. Engr.	Fore-man
Hand mixing using one mixing board.........	3.33*	..	..	0.33
Hand mixing using two mixing boards........	3**	..	..	0.25
Using one sack (6-S) mixer and wheeling to place...........	2.5†	..	..	0.17
Using two sack (11-S) mixer and wheeling to place...........	2	0.125	..	0.125
Using 16-S mixer and wheeling to place......	2	0.125	..	0.125
Using 28-S mixer and wheeling to place......	2	0.083	..	0.08
Using conveyors, batcher-plant, hoist and wheeling to place...........	1.25	0.06	0.06	0.06

Labor Placing One Cu. Yd. of Ready-Mixed Concrete for Concrete Foundations

Method and Quantity Placed per 8-Hr. Day	Hours per Cubic Yard		
	Common Labor	Hoisting Engineer	Fore- man
Placing concrete directly in forms, 1 truck servicing job, 50 to 70 cu. yds	0.35-0.50	....	0.13
Placing concrete directly in forms, 2 trucks servicing job, 100 to 150 cu. yds	0.27-0.40	....	0.08
Using wheelbarrows, 50 to 70 cu. yds	1.25-1.50	0.17*	0.17**
Using wheelbarrows, 100 cu. yds. and over	1.00-1.25	0.13*	0.13**
Using concrete buggies, 50 to 70 cu. yds	1.00-1.25	0.17*	0.17**
Using concrete buggies, 100 cu. yds and over	0.90-1.10	0.13*	0.13**
Using 13 cu. ft. power concrete buggies	0.65-0.75	0.13*	0.13**

*Where the concrete is hoisted and it is necessary to use a hoisting engineer, add as given in table.
**If union regulations require a mason or cement finisher foreman in charge of crew, add as given in table.

Labor Placing One Cu. Yd. of Ready-Mixed Concrete for Reinforced Floors, Beams, Girders, Columns, Etc.

Method of Handling and Placing the Concrete	Hours per Cubic Yard		
	Common Labor	Hoisting Engineer	Fore- man
Hoisting concrete in wheelbarrows on a material hoist and wheeling into place.	2.3	.11*	.11*
Hoisting concrete in concrete buggies on a material hoist and wheeling into place.	1.1	.08*	.08*
Thin floors 2 to 2½-in. thick. Hoisting concrete in wheelbarrows on a material hoist and wheeling into place.	2.8	.11*	.11*

	Hours per Cubic Yard		
Thin floors 2 to 2½-in. thick. Hoisting concrete in concrete buggies on a material hoist and wheeling into place	Common Labor	Hoisting Engineer	Foreman
	1.6	.08*	.08*
Hoisting concrete in a concrete bucket, discharging into a floor hopper and wheeling into place, using wheelbarrows......................	1.2	.05*	.05*
Hoisting concrete in a concrete bucket, discharging into a floor hopper and wheeling into place, using concrete buggies..................................	0.8	.04*	.04*
If power buggies are used for transporting concrete from hopper to place of deposit, deduct per cu. yd	0.25-0.33	...	...
Mobile crane with bucket on ground, hoisting concrete to floor or forms and depositing directly into forms.................................	0.5	...	.04*
Add for rental of mobile crane and operator.....................................	...	.04*	...
For thin concrete floors, 2 to 2½-in. thick, add to the above costs	0.5	...	...

*Time of hoisting engineer and foreman will vary according to the number of cubic yards of concrete placed per day.

PLACING CONCRETE BY PIPELINE

Concrete pumps can handle efficiently concrete of any slump from ½" to 7" but the most dependable slump for general conditions is around 3" to 5". However, plastic concrete of about 6" slump can generally be pumped the maximum distance and height. The rate of pumping depends considerably upon the consistency of the concrete.

Placing Concrete on a Commercial Building.—Assume a 5-story steel frame commercial building, 140 by 250 ft. having floors of metal pan construction, with pans from 6" to 14" deep and 2½" of concrete over the pans, the entire job containing approximately 5,200 cu. yds. of concrete.

Assume the concrete is ready-mixed, using ¾" gravel aggregate and having a 5" slump.

The concrete is delivered to the job and deposited directly into the remixing hopper on the pumper and pumped to the various floors.

The labor per cu. yd. cost as follows:

	Hour	Rate	Total	Rate	Total
Pump operator	0.07	$....	$....	$17.95	$1.26
Hopper man	.07	....	....	12.54	.88
Labor vibrating, pipeline, etc.	.49	....	....	12.54	6.15
Foreman	.07	....	....	13.04	.91
Cost per cu. yd.			$....		$9.20

Average rate of pour 15 cu. yds. per hr. Maximum rate of pour 19 cu. yds. per hr.

The above costs do not include the cost of the pumper, depreciation, parts and replacements, gas and oil, labor installing and removing equipment, etc.

CAST IN PLACE CONCRETE

Concrete floors, sidewalks, and lightweight concrete floor fill are estimated by the sq. ft., taking the number of sq. ft. of any given thickness, as 2", 3", 4", etc. to obtain the total cubic yards for placement.

Number of Sq. Ft. of Concrete Floor of Any Thickness Obtainable From One Cu. Yd. of Concrete

Thickness Inches	No. Sq. Ft.	Thickness Inches	No. Sq. Ft.	Thickness Inches	No. Sq. Ft.	Thickness Inches	No. Sq. Ft.
1	324	4	81	7	46	10	32
1¼	259	4¼	76	7¼	44	10¼	31
1½	216	4½	72	7½	43	10½	31
1¾	185	4¾	68	7¾	42	10¾	30
2	162	5	65	8	40	11	29½
2¼	144	5¼	62	8¼	39	11¼	29
2½	130	5½	59	8½	38	11½	28
2¾	118	5¾	56	8¾	37	11¾	27½
3	108	6	54	9	36	12	27
3¼	100	6¼	52	9¼	35	12¼	26½
3½	93	6½	50	9½	34	12½	26
3¾	86	6¾	48	9¾	33	12¾	25½

Labor Placing Ready-Mixed Concrete for Floors.—In order to place ready-mixed concrete for floors with the most economy it is necessary to have a crew large enough to handle the concrete quickly to eliminate waiting time charges by the material company and the supply of mixed concrete must be steady and of sufficient quantity to keep the crew busy.

A crew of 14 or 15 laborers, using buggies to wheel concrete, should place 22 to 27 cu. yds. of ready-mixed concrete per hr. at the following labor cost per cu. yd.:

	Hours	Rate	Total	Rate	Total
Labor	0.6	$....	$....	$12.54	$7.53

Labor Applying A Finish on Reinforced Concrete Floors.
—Architects and engineers specify the floor must be finished at the same time the concrete is placed, and in some instances it necessitates a large amount of overtime for finishers.

Where the gang places concrete right up until quitting time, it is a foregone conclusion that the finishers will have to work far into the night.

The item of overtime should be carefully considered by the estimator for as a usual thing the cement mason starts work when all other trades have quit for the day.

While under favorable conditions, some finishers may finish 175 to 200 sq. ft. of floor an hour on certain classes of work, the general average for a reinforced concrete job will vary from 70 to 80 sq. ft. an hr. or 560 to 640 sq. ft. per 8-hr. day.

The labor costs given are for straight time but the estimator should bear in mind that overtime costs time and a half or double time.

The labor per 100 sq. ft. of floor should cost as follows:

	Hours	Rate	Total	Rate	Total
Cement mason	1.3	$....	$....	$16.62	$21.61
Cost per sq. ft			....		.22

Finishing Concrete Floors by Machine.
—Over the years rapid strides have been made in the finishing of concrete floors by power operated machines. These machines are used for screeding, floating and troweling the finish floor.

For floating, the machine is equipped with 12 gauge steel trowels, which revolve on a 46" diameter. It usually requires 3 cement masons; 1 finisher to operate the machine and 2 finishers touching up corners, columns, edging, jointing, etc. On a well prepared slab, one finishing crew can float 4,000 sq. ft. of floor an hr. but this is not usually maintained due to job delays, etc.

After the floor has been floated, the heavy gauge steel floating trowels are removed and replaced with steel "finishing" trowels. The cement mason then guides the rotating, adjustable pitch trowels over the slab until a smooth, level surface is obtained.

Using the larger machines, and assuming a "four time over" was necessary to complete the job, it would require 4 hrs. of machine operation and assuming a ¼ to 1-hr. waiting period was necessary between each operation, it would require a total of 6¼ to 7 hrs. time to finish 4,000 sq. ft. of floor.

After the floor has been screeded, the labor floating and

troweling (4 times over), 4,000 sq. ft. of cement finish floor should cost as follows:

	Hours	Rate	Total	Rate	Total
Cement mason	32	$....	$....	$16.62	$531.84
Cost per sq. ft			....		0.133

*Add Machine Rental Charges

Lightweight Concrete, Insulating Concrete of Pumice, Perlite, Vermiculite and Other Lightweight Aggregates

Lightweight concrete of crushed slag, foamed slag, haydite, pumice, perlite and vermiculite are used for floor and roof fill, structural roofs, floors in cold storage rooms, etc.

A comparison of concrete aggregates is as follows:

Type of Aggregate	Aggregate Weight per cubic foot Pounds	Weight per Cu. Ft. of Concrete Using Aggregate, Pounds
Gravel	120	150
Sand	90-100	150
Crushed stone	100	145

	Aggregate Wt. per Cu. Ft. Lbs.	Concrete Wt. per Cu. Ft. Lbs.
Crushed Bank Slag	80	100-130
Haydite	40-60	100-120
Foamed slag	40-60	90-100
Cinders	40-50 (plus sand)	110-115
Pumice	30-60	60- 90
Diatomite	28-40	55- 70
Perlite	6-16	20- 50
Vermiculite	6-10	20- 40
Waylite	40-60	90-100

CEMENT FLOOR HARDENERS

There are many preparations on the market for waterproofing, dustproofing and wearproofing cement finish floors. These preparations fall into two classifications, integral treatments in which the entire slab or topping is treated and surface treatments in which the surface is specially prepared to become waterproof, dustproof and wear resistant.

Integral Treatments.—Integral treatments consist of powdered chemicals that are mixed with the cement before sand and water are added and become an integral part of the topping mix throughout its entire thickness. This type of treatment is said to densify and harden the topping and is generally recommended for areas subjected to light pedestrian traffic.

Surface Treatments.—There are two types of surface treatments for cement finish floors, dust coat applications which must take place as floors are being finished and liquid chemical applications which may be done at any time after the floor finish is set hard enough for foot traffic.

Liquid Chemical Floor Treatments (Hardeners)

As concrete wears down it is ground into dust and this condition has had an especially detrimental effect in mills and factories where the dust from the floors has been carried into the bearings of machinery, motors, engines, etc., causing considerable damage and expense in the upkeep of the equipment.

To overcome this condition, there have been a number of chemicals in liquid form placed on the market that create a chemical action in the cement and lime in the floors and cause a complete transformation of the lime so that it solidifies the entire concrete aggregate into a hard, flint-like, homogenous mass that prevents dusting and wearing of the floors.

These liquid chemical compounds are sprayed over the floor until all of the pores in the surface are filled. It is usually necessary to treat the floors two or three times with the liquid to secure satisfactory results. When using these preparations, it is necessary to treat the floors at stated intervals, as the chemical action they create on the cement and lime is not permanent, although some manufacturers claim their product is permanent.

When using these liquid chemicals care must be exercised to see that the floor is absolutely clean and free from all grease, dust, dirt, oil, or other foreign matter.

Concrete Accelerators and Densifiers

There are numerous preparations on the market for controlling the set of concrete, increasing its early strength, densifying and waterproofing the concrete mass, and for preventing freezing during winter weather by lowering the freezing point of the mixing water.

Anti-Hydro Accelerator and Anti-Freeze. (Anti-Hydro Co., Newark, N. J.)—Can be mixed with the water or added to the wet mix. As a general rule, the standard proportion of 1.5 gals. of Anti-Hydro per cu. yd. of concrete, with reduced water content to compensate for increased slump, will give protection down to 23° F. on concrete deposited in forms except in the most exposed locations.

For cold weather work, be sure that sand and aggregate are free of ice.

To protect concrete or mortar against freezing, use the following table for anticipated outside air temperatures as follows:

32° to 25° F. use 1 part Anti-Hydro to 12 parts water.

25° to 15° F. use 1 part Anti-Hydro to 10 parts water.

15° or less, mechanical heat is necessary.

Calcium Chloride.—A white, dry, flaky chemical which, when added to the mixing water accelerates the initial hardening of concrete; at the same time densifies the concrete, making it more waterproof; reduces the freezing point of water, thereby aiding the fresh concrete to resist freezing in low temperatures.

Flake calcium chloride may be added to the mix in either dry form (as it comes from the package) or in the form of a solution. The solution is made up of flake calcium and water in proportions that will result in a liquid each quart of which will be equivalent to one lb. of 77–80% flake calcium chloride. One quart of standard solution can be substituted wherever one lb. of flake material is specified.

Calcium chloride is always added to the mix in terms of "pounds per sack of portland cement." The quantities recommended vary from one pound per sack of cement for use when the temperature is above 90° F. to as much as four pounds per sack of cement during the lower range of temperatures. Maximum efficiency is obtained in the use of up to two pounds the proportional reaction of quantities in excess of this amount being less marked. The use of quantities in excess of four pounds per sack of cement is not recommended at any time.

When dry flake calcium chloride is used it should be added to and with the aggregate and not the cement. When solution is used it should be introduced with the mixing water (which should in all cases be reduced by at least the amount of calcium chloride solution used).

The following quantities are recommended per sack of portland cement:

For temperatures above 90° F. .... 1 lb.
For temperatures 80° to 90° F. .... 1.5 lbs.
For temperatures 32° to 80° F. .... 2 lbs.
For temperatures below 32° F. .... 2 to 4 lbs.,

BONDING OLD AND NEW CONCRETE

The problem of bonding new concrete to old so the new concrete will remain permanently in place, is one that has received considerable attention among users of concrete. Unless the utmost care is exercised in preparing the old surface, the new concrete will invariably come loose where it is joined to the old concrete.

Different methods have been used to overcome this fault, but the basic condition of all of them is that the old surface must be thoroughly cleaned and washed, and the old aggregate must be exposed, which means that the thin film of cement that covers the surface of the concrete must be removed. One of the methods of preparing the old surface consists of hacking or picking it, and then washing or turning a steam hose under pressure to remove all dust and dirt from the old floor so as to present a perfectly clean surface.

Hacking and Chipping Old Concrete.—The cost of hacking and chipping old concrete surfaces in preparation for bond-

ing new concrete to old, is a variable item, depending upon the
hardness of the old concrete and the condition of the surface.
Under average conditions, a laborer using a pick, should hack
and roughen 175 to 200 sq. ft. of "green" concrete floor per
8-hr. day, at the following cost per 100 sq. ft.:

	Hours	Rate	Total	Rate	Total
Labor	4	$....	$....	$12.54	$50.16
Cost per sq. ft			....		.50

Where the concrete floors are of old concrete, a man using
hand tools will do well to hack and roughen 8 to 10 sq. ft. an
hour, and even when a compressor and jack-hammer or bush-
hammering tool is used it is often possible to hack and roughen
only 20 to 25 sq. ft. an hr. and the cost per 100 sq. ft. should
average as follows:

	Hours	Rate	Total	Rate	Total
Compressor expense	2.25	$....	$....	$ 7.50	$16.88
Labor on jack-hammer	4.50	....	....	12.54	56.43
Cost per 100 sq. ft			$....		$73.31
Cost per sq. ft			....		.73

Weld-Crete.*—A product that permanently bonds new
concrete to old. It also bonds new concrete to many
other materials, such as brick, stucco, stone, etc.

For successful bonding of new concrete to old, the old concrete surface must be structurally sound and free from dust,
dirt, loose material, grease, oil, wax, water soluble coatings, etc.
It is not necessary to chip, bush hammer or roughen the old
surface in any way.

Weld-Crete may be used on either interior or exterior surfaces and may be applied over "green" concrete and damp or
dry surfaces. May be applied with a brush, roller or spray.
Spraying is most efficient and should be done with heavy industrial spray equipment.

Coverage, when sprayed, from 200 to 300 sq. ft. per gal.
When brushed or rolled, coverage is somewhat less.

03400   PRECAST CONCRETE

FLEXICORE FLOOR AND ROOF SLABS

Prestressed Flexicore concrete floor and roof slabs designed
in accordance with ACI 318 Building Code are precast in cen-
tral plants with patented equipment. The cross sections and
lengths of these units are indicated in Table I. Filler width
slabs are also available.

	TABLE 1 SPAN-FT.	LBS. PER SQ. FT. (Based on 150 Lb. Concrete)
6"x24"	15-25	43
8"x24"	20-33	57
10"x20"	28-40	61
10"x24"	28-40	72
12"x24"	35-50	79

These units are manufactured in a rigid steel form made sufficiently strong to resist the pretensioning force applied by the seven wire strand reinforcing steel. Specially constructed rubber tubes are inflated to form circular voids so as to constitute approximately 50% of the cross section. The concrete is thoroughly vibrated to assure maximum denseness and strength.

The slabs may be cast of concrete made of gravel, crushed stone or lightweight aggregate. When lightweight aggregate is used, the unit weight is approximately 25% less. They are cured in a heated kiln.

Erection.—Flexicore units are usually delivered to the job by truck. They are hoisted from the truck to the floor location and often placed directly into their final position by the crane. After the slabs are placed side by side, they are aligned and levelled. The keyways in the sides of the slab are then filled with a grout mixed in a ratio of one to three. The erection crew usually consists of six men, including a crane operator, an oiler, a foreman and three laborers. A crew will unload, erect and grout 2800 sq. ft. per day on smaller jobs to approximately 6000 sq. ft. on some larger jobs.

CHAPTER 4

MASONRY

CSI   DIVISION 4

04100   MORTAR

The following mortars are based on proportion set forth in ASTM specifications. These mortars were designated Type A-1, A-2, B and C prior to 1954.

Type M mortar is a high strength mortar used primarily in foundation masonry, retaining walls, walks, sewers and manholes. Its proportions and costs for a cement-lime mixture are:

Proportions and costs for a type M mortar using masonry cement are:

```
1 bag of portland cement ............... $3.75
1 bag Type II masonry cement ......  3.25
6 cu. ft. sand @ $0.22 per
    cu. ft. ..............................  1.32
                                         $8.82 or $1.39 per cu. ft.
```

Type S mortar also has a reasonably high compressive strength and develops maximum tensile bond strength between brick and cement-lime mortars. It is recommended for use in reinforced masonry and where flexural strengths are required, such as cavity walls exposed to strong winds, and for maximum bonding power, such as for ceramic veneers.

Costs and proportions for cement masonry Type S Mortar are:

```
1/2 bag portland cement ............... $3.75
1 bag masonry cement ....................  3.25
41/2 cu. ft. sand @ $0.22
    cu. ft. ...............................   .99
                                         $7.99 or $1.78 per cu. ft.
```

Type N mortar is a medium strength mortar most generally used in exposed masonry above grade.

Proportions and cost of Type N masonry cement mortar is:

```
1 cu. ft. masonry cement ............... $3.25
3 cu. ft. sand @ $0.22
    cu. ft. ...............................   .66
                                         $3.91 or $1.30 per cu. ft.
```

Cu. Ft. of Mortar Required to Lay 1,000 Face Brick

Based on standard size brick having 1/4" to 3/8" end joints and "bed" joints as follows:

			Width of Mortar Joints in Inches				
1/8"	1/4"	3/8"	1/2"	5/8"	3/4"	7/8"	1"
4	7	9	12	14	16	18	20

For Dutch, English and Flemish Bond, add about 10 percent to the above quantities on account of additional head joints.

Cubic Feet of Mortar Required per Thousand Brick

For various thicknesses of walls and joints. No allowance for waste

Joint Thickness	4" Wall	8" Wall	12" Wall	16" Wall	20" Wall	24" Wall
1/8"	2.9	5.6	6.5	7.1	7.3	7.5
1/4"	5.7	8.7	9.7	10.2	10.5	10.7
3/8"	8.7	11.8	12.9	13.4	13.7	14.0
1/2"	11.7	15.0	16.2	16.8	17.1	17.3
5/8"	14.8	18.3	19.5	20.1	20.5	20.7
3/4"	17.9	21.7	23.0	23.6	24.0	24.2
7/8"	21.1	25.1	26.5	27.1	27.5	27.8
1"	24.4	28.6	30.1	30.8	31.2	31.5

Quantity of Mortar Required to Lay 1,000 Concrete Brick of Various Sizes

Width of Bed Joint	1 Brick or 4" Wall	2 Bricks or 8" Wall	2 Bricks or 8" Backup	3 Bricks or 12" Wall
	Modular Size Brick, 2¼"x3⅝"x7⅝"			
5/12"	9.64 cu. ft.	12.19 cu. ft.	14.73 cu. ft.	13.01 cu. ft.
	Jumbo Brick, 3⅝"x3⅝"x7⅝"			
3/8"	10.06 cu. ft.	13.88 cu. ft.	17.70 cu. ft.	15.15 cu. ft.
	Double Brick, 4⅞"x3⅝"x7⅝"			
11/24"	12.68 cu. ft.	17.77 cu. ft.	22.86 cu. ft.	19.47 cu. ft.
	Roman Brick, 1⅝"x3⅝"x11⅝"			
3/8"	11.79 cu. ft.	14.65 cu. ft.	....	....
	Roman Brick, 1⅝"x3⅝"x15⅝"			
3/8"	15.25 cu. ft.	19.07 cu. ft.	....	....

Mortar Required for Setting Stone.—The quantity of mortar required for stone setting will vary with the size of the stone, width of bed, etc., but as a general thing it will require 4 to 5 cu. ft. of mortar per 100 cu. ft. of stone.

If the stone is to be back-plastered, the quantity of mortar will vary with the thickness of the plaster coat but as most ashlar is 4" to 8" thick, it will require the following cu. ft. of mortar per 100 cu. ft. of stone:

Plaster Thickness	Thickness of Stone in Inches		
	4"	6"	8"
¼"	6½	4¼	3¼
½"	13	8½	6½
¾"	19	13	9½

It is customary to use white non-staining portland cement for setting and back-plastering limestone or other porous stone, as this prevents stains from appearing on the face of the stone. Never use ordinary portland cement for setting limestone.

04200　UNIT MASONRY
04210　BRICK

The Structural Clay Products Institute separates brick walls into 3 general categories: conventional, bonded walls; drainage type walls; and barrier type walls.

The drainage type walls include brick veneer, the "SCR brick" wall, the cavity type wall, and the masonry bonded hollow wall. Drainage type walls are recommended for walls subjected to the most severe exposures. The success of this type of wall, of course, depends on the care taken to provide continuous means for water to escape.

The barrier type walls include metal tied walls and reinforced brick masonry. The success of these walls will depend on a solidly filled collar joint. The SCR presently recommends using an 8" metal tied wall in preference to an 8" bonded wall for areas subjected to moderate exposure. Metal tied walls adapt well to brick and block backup.

Bonded walls are still widely used and can be more interesting in design. Presently, there is renewed interest in interior masonry bearing walls, applied in some cases to high rise buildings. In Europe, buildings as high as 16 stories are being built with the full load of the structure carried on 8" walls. It is most often encountered in this country in apartment and dormitory type buildings of 3 to 4 stories high with many fixed interior partitions. In this type of construction, the compressive strength of the brick and mortar is of great importance.

Number of Standard Brick (8"x2¼"x3¾") Required for One Sq. Ft. of Brick Wall of any Thickness

Thickness of wall	Number of Brick Thick	Vertical or End Mortar Joints figured as ¼" wide Width of Horizontal or "Bed" Mortar Joints					
		⅛"	¼"	⅜"	½"	⅝"	¾"
4" or 4½"	1	7.33	7	6.67	6.33	6.08	5.8
8" or 9"	2	14.67	14	13.33	12.67	12.17	11.6

*12" or 13"	3	22.00	21	20.00	19.00	18.25	17.4
16" or 17"	4	29.33	28	26.67	25.33	24.33	23.2
20" or 21"	5	36.67	35	33.33	31.67	30.42	29.0
24" or 25"	6	44.00	42	40.00	38.00	36.50	34.8

*Use this column for computing the number of brick required per cu. ft. of wall with any width mortar joint.

Modular Brick Walls Without Headers

Nominal Brick Size in Inches			Brick Per Sq Ft of Wall	Cubic Feet of Mortar Per 100 Sq Ft of Wall			Cubic Feet of Mortar Per 1000 Brick		
				Joint Thickness			Joint Thickness		
h	t	l		¼ in.	⅜ in.	½ in.	¼ in.	⅜ in.	½ in.
2 2/3	x4	x 8	6.750	3.81	5.47	6.95	5.65	8.10	10.30
3 1/5	x4	x 8	5.625	3.34	4.79	6.10	5.94	8.52	10.84
4	x4	x 8	4.500	—	4.12	5.24	—	9.15	11.65
5 1/3	x4	x 8	3.375	—	3.44	4.34	—	10.19	12.87
2	x4	x12	6.000	—	6.43	8.20	—	10.72	13.67
2 2/3	x4	x12	4.500	3.52	5.06	6.46	7.82	11.24	14.35
3	x4	x12	4.000	—	4.60	5.87	—	11.51	14.68
3 1/5	x4	x12	3.750	3.04	4.37	5.58	8.11	11.66	14.89
4	x4	x12	3.000	2.56	3.69	4.71	8.54	12.29	15.70
5 1/3	x6	x12	2.250	—	3.00	3.84	—	13.34	17.05
2 2/3	x6	x12	4.500	—	7.85	10.15	—	17.45	22.55
3 1/5	x6	x12	3.750	—	6.79	8.77	—	18.10	23.39
4	x6	x12	3.000	—	5.72	7.40	—	19.07	24.67

The above quantities include mortar for bed and vertical joints only. The following table gives allowances for backing mortar or collar joints.

Cubic Feet of Mortar Per 100 Sq Ft of Wall		
¼ -in. Joint	⅜ -in. Joint	½ -in. Joint
2.08	3.13	4.17

Note:

$$\frac{\text{Cubic feet}}{\text{per 1000 units}} = \frac{10 \times \text{cubic feet per 100 sq ft of wall}}{\text{number of units per square foot of wall}}$$

Labor Production For Laying Building Brick

Class of Work	Mortar Joints Style	Average Number Brick Laid per 8-hr. Day	Aver. Hrs. 1,000 Brick	
			Mason Hours	Labor Hours
8" walls, 1-story bungalows, garages, two-flat buildings, residences, etc.	Cut	750-800	10.5	9.0
	Struck	700-750	11.0	9.0
12" walls, ordinary construction, apartment buildings, houses, garages, factories, stores and apartment buildings, schools, etc.	Cut	950-1,050	8.0	8.0
	Struck	925-975	8.5	8.0
	Cut	825-900	9.3	8.0
	Struck	750-850	10.0	8.0
Hodding brick to second story, add			2.0	
16" walls, heavy warehouse, factory and industrial work. Straight	Cut	1,100-1,225	7.0	8.0
	Struck	1,000-1,125	7.5	8.0
	Cut	950-1,050	8.0	8.0
	Struck	900-1,000	8.4	8.0
Backing-up face brick, cut stone, terra cotta on ordinary wall bearing buildings, figure same as given above for 8" and 12" walls				

WINTER WORK. Heating brick, mortar, attending salamanders, removing snow and ice, add extra for this work.

*HOISTING. Add about ¼-hr. hoisting engineer time per 1,000 for reasonable size jobs using 35,000 or more brick per day; ½-hr. per 1,000 for jobs using 15,000 to 20,000 brick per day; and ⅓-hr. per 1,000 for jobs using 25,000 to 30,000 brick per day.

Labor Production For Laying Building Brick—Cont.

Class of Work	Mortar Joints Style	Average Number Brick Laid per 8-hr. Day	Aver. Hrs. 1,000 Brick	
			Mason Hours	Labor Hours
Backing-up face brick, terra cotta on steel or concrete skeleton frame buildings. First Grade workmanship. Walls 8" to 12" thick.	Cut	825-900	9.3	9.0
	Struck	750-850	10.0	9.0
	Cut	725-800	10.5	9.0
	Struck	700-775	10.8	9.0
Public buildings. First Grade workmanship. Schools, college and university bldgs. Courthouses, state capitols, public libraries, etc. 12" to 20" walls.	Cut	850-950	9.0	8.0
	Struck	800-900	9.4	8.0
	Cut	750-850	10.0	8.0
	Struck	700-800	10.6	8.0
Powerhouses and other structures, having high walls 12" to 20" thick, without intermediate floors.	Cut	1,100-1,200	7.0	9.0
	Struck	1,000-1,100	7.6	9.0
	Cut	925-1,050	8.0	9.0
	Struck	875-950	8.8	9.0
Shoved joints. Brick laid in full shoved joints with all vertical joints slushed full of mortar, add to all of the above			1.5	
Basement foundation walls, paving brick.	Cut	900-1,000	8.5	8.5
	Struck	850-950	9.0	9.0

WINTER WORK. Heating brick, mortar, attending salamanders, removing snow and ice, add extra for this work.
*HOISTING. Add about ¼-hr. hoisting engineer time per 1,000 for reasonable size jobs using 35,000 or more brick per day; ½-hr. per 1,000 for jobs using 25,000 to 30,000 brick per day; and ⅓-hr. per 1,000 for jobs using 15,000 to 20,000 brick per day; and ⅓-hr. per 1,000 using 13,000 to 20,000 brick per day.

Labor Production For Laying Building Brick—Cont.

Class of Work	Mortar Joints Style	Average Number Brick Laid per 8-hr. Day	Aver. Hrs. 1,000 Brick	
			Mason Hours	Labor Hours
Building brick foundation walls 8" to 12" thick, ordinary workmanship.	Cut	1,000–1,100	7.6	8.0
	Struck	900–975	8.5	8.0
	Cut	900–975	8.5	8.0
	Struck	825–900	9.3	8.0
Chimneys and stacks, building brick, 1'-4" to 2'-0" sq, 15'-0" above roof, 4" to 8" walls. Hodding brick, extra.	Struck	500–550	16.0	16.0
	......			3.0
Large chimney and stacks, 3'-0" to 4'-0" Sq. 15' to 30' high above roof, 8" to 12" walls.	Struck	550–600	14.0	16.0
Large brick stacks 100' to 150' high. walls from 1'-8" at base to 12" top. Inside put-log scaffold. Outside scaffold extra.	Struck	650–750	11.5	14.0
Bricking-in boilers, fire-boxes, etc.	Struck	600–650	12.5	10.0

WINTER WORK. Heating brick, mortar, attending salamanders, removing snow and ice, add extra for this work.

*HOISTING. Add about ¾-hr. hoisting engineer time per 1,000 for reasonable size jobs using 35,000 or more brick per day; ½-hr. per 1,000 for jobs using 15,000 to 20,000 brick per day; and ⅓-hr. per 1,000 for jobs using 25,000 to 30,000 brick per day.

Twin Brick

Twin brick is a double size brick that contains the face brick and a 4" backing brick in one unit, size 8"x8"x2¼".

It is made with a hollow core so that the twin brick weighs only 7½ lbs. and can be picked up and laid in a one-hand operation. It is designed so the wall has no continuous mortar joint from outer to inner wall faces, thereby tending to avoid moisture penetration.

It is intended primarily for residential work having 8" walls that are to be furred and plastered. Extra face brick must be used at openings and corners to insure proper bond.

It requires about 25 cu. ft. of mortar per 1,000 units.

On four jobs on which these brick were used, requiring 50,000 units, a mason laid 60 twin brick an hour or 480 per 8 hr. day, at the following labor cost per 1,000 units:

	Hours	Rate	Total	Rate	Total
Mason	17	$....	$....	$18.91	$321.47
Labor	11.5			14.95	171.93
Cost per 1,000 units			$....		$493.40

Brick Floors and Steps

Face Brick Floors Laid in Basket Weave Pattern.—Brick floors laid in square or basket weave pattern do not require any cutting and are much easier laid than herringbone designs.

On work of this kind, a mason should lay 180 to 225 brick per 8-hr. day, at the following labor cost per 1,000:

	Hours	Rate	Total	Rate	Total
Mason	40	$....	$....	$18.91	$ 756.40
Labor	20	....	....	14.95	299.00
Cost per 1,000 brick			$....		$1,055.40

Face Brick Floors Laid in Herringbone Pattern.—Brick floors are usually laid in portland cement mortar and when laid in herringbone pattern, it requires considerable time cutting and fitting the small pieces of brick at the edge or border of the pattern, as shown in the illustration.

A mason should lay 125 to 150 brick per 8-hr. day, at the following labor cost per 1,000:

	Hours	Rate	Total	Rate	Total
Mason	56	$....	$....	$18.91	$1,058.96
Labor	28	....	....	14.95	418.60
Cost per 1,000 brick			$....		$1,477.56

If there is only one mason on the job it will require a helper for each mason.

Brick Steps.—When laying brick steps in portland cement mortar, a mason should lay 125 to 150 brick per 8-hr. day, at the following labor cost per 1000:

	Hours	Rate	Total	Rate	Total
Mason	56	$....	$....	$18.91	$1,058.96
Labor	28	....	....	14.95	418.60
Cost per 1,000 brick			$....		$1,477.56

Turning Face Brick Segmental Arches Over Doors and Windows.—Where segmental brick arches are laid over wood centers for 3'-0" wide door and window openings, it will require about 4 hrs. mason time at the following labor cost per arch:

	Hours	Rate	Total	Rate	Total
Mason	4	$....	$....	$18.91	$75.64

For a 5'-0" arch, figure about 6.5 hrs. mason time, as follows:

	Hours	Rate	Total	Rate	Total
Mason	6.5	$....	$....	$18.91	$122.92

If the brick are cut or chipped to a radius, add cost of chipping as given below.

Laying Face Brick in Flat or "Jack" Arches.—When laying face brick in flat or "Jack" arches a mason should complete one arch up to 3'-6" wide in about 5 hrs. at the following labor cost:

	Hours	Rate	Total	Rate	Total
Mason	5	$....	$....	$18.91	$94.55

For a 5'-0" "Jack" arch, figure about 7 1/2 hrs. mason time, as follows:

	Hours	Rate	Total	Rate	Total
Mason	7.5	$....	$....	$18.91	$141.83

The above costs do not include chipping or rubbing brick to a radius. If this is necessary, add cost of chipping as given below.

Cleaning Face Brick Work

The cost of cleaning face brick work will vary with the kind of brick, as it is much easier to clean a smooth face vitrified brick than one having a rough texture.

Where smooth face brick are used and it is not necessary to do any pointing but merely wash the wall with muriatic acid and water, an experienced mechanic should clean 5,000 to 5,500 brick (750 to 825 sq. ft. of wall) per 8-hr. day, at the following labor cost per 1,000 brick, or approximately 150 sq. ft. of wall:

	Hours	Rate	Total	Rate	Total
Mechanic	1.6	$....	$....	$18.91	$30.26
Helper	0.8	....	....	12.54	10.03
Cost per 1,000 brick			$....		$40.29
Cost per sq. ft. of wall			....		.27

When cleaning and washing rough textured brick, the same crew should clean and wash 3,500 to 4,000 brick (525 to 600 sq. ft. of wall) per 8-hr. day, at the following labor cost per 1,000 brick or per 150 sq. ft. of wall:

	Hour	Rate	Total	Rate	Total
Mechanic	2.2	$....	$....	$18.91	$41.60
Helper	1.1	....	....	12.54	13.80
Cost per 1,000 brick			$....		$55.40
Cost per sq. ft. of wall			....		.37

Brick Fireplaces, Mantels and Hearths

A mason should complete a plain brick fireplace 5'-0" to 6'-0" wide and 4'-0" to 5'-0" high, requiring 200 to 225 brick, not including hearth, in about 10 hrs. at the following labor cost:

	Hours	Rate	Total	Rate	Total
Mason	10	$....	$....	$18.91	$189.10
Labor	5	....	....	14.95	74.75
Cost per fireplace			$....		$263.85

On larger and more elaborate brick fireplaces and mantels having raked or rodded mortar joints, a mason should lay about 100 to 125 brick per 8-hr. day, at the following labor cost per 1,000:

	Hours	Rate	Total	Rate	Total
Mason	70	$....	$....	$18.91	$1,323.70
Labor	35	....	....	14.95	523.25
Cost per 1,000 brick			$....		$1,846.95

If there is only one mason working on the fireplace it will require one laborer with each mason.

FIRE BRICK WORK

Labor Lining Brick Chimneys and Stacks with Fire Brick.—When lining small brick chimneys and stacks from 2'-

0" to 3'-0" square, laid in fire clay, a mason should lay 450 to 525 brick per 8-hr. day, at the following labor cost per 1,000:

	Hours	Rate	Total	Rate	Total
Mason	16	$....	$....	$18.91	$302.56
Labor	12	....	....	14.95	179.40
Cost per 1,000 brick			$....		$481.96

On large bricks stacks having an inside diameter of 4'-0" to 6'-0", a mason should lay 600 to 750 fire brick per 8-hr. day, at the following labor cost per 1,000:

	Hours	Rate	Total	Rate	Total
Mason	11.8	$....	$....	$18.91	$223.14
Labor	10.0	....	....	14.95	149.50
Cost per 1,000 brick					$372.64

Labor Laying Fire Brick in Fireboxes, Breechings, Etc.—When lining fireboxes, breechings, furnaces, etc., with fire brick, requiring arch and radial brick, special shapes, etc., a mason should lay 175 to 225 brick per 8-hr. day, at the following labor cost per 1,000:

	Hours	Rate	Total	Rate	Total
Mason	40	$....	$....	$18.91	$756.40
Labor	40	....	....	14.95	598.00
Cost per 1,000 brick					$1,354.40

Fire Clay Tile Flue Lining

Flue lining is estimated by the lin. ft. It is furnished in 2'-0" lengths in the following sizes:

Size of Flue Lining	No. Lin. ft. Set per 8-Hr. Day	Hrs. Required to Set 100 Lin. Ft. Mason	Labor
4"x 8"	165–180	4.7	4.7
4"x12"	130–150	5.7	5.7
8"x 8"	130–150	5.7	5.7
8"x12"	105–125	7.0	7.0
12"x12"	85–100	8.7	8.7
8"x18"	75– 90	9.7	9.7
12"x18"	70– 85	10.2	10.2
18"x18"	60– 70	12.3	12.3
20"x20"	55– 65	13.3	13.3
20"x24"	50– 60	14.0	14.0
24"x24"	45– 55	16.0	16.0
20" round	55– 65	13.3	13.3

Brick Catch Basins and Manholes

When laying building or sewer brick in catch basins, manholes, etc., from 3'-0" to 5'-0" in diameter and having 8" to 12" walls, a mason should lay 600 to 750 brick per 8-hr. day, at the following labor cost per 1,000 :

	Hours	Rate	Total	Rate	Total
Mason	11.5	$....	$....	$18.91	$217.47
Labor	11.5	....	....	14.95	171.93
Cost per 1,000 brick			$....		$389.40

04220   CONCRETE UNIT MASONRY

Sizes, Weights and Quantities of Load-Bearing Concrete Blocks and Tile

Actual Size and Description of Units	Wall Thickness	Approx. Weight Lbs. Hvy. Wt. Units	Approx. Weight Lbs. Light Wt. Units	No. Units 100 Sq. Ft. of Wall	Cu. Ft. Mortar Per 100 Sq. Ft. of Wall*
3⅝"x4⅞"x11⅝"	4"	11-13	8-10	240	8.0
5⅝"x4⅞"x11⅝"	6"	17-19	12-14	240	8.5
7⅝"x4⅞"x11⅝"	8"	22-24	14-16	240	9.0
3⅝"x7⅝"x11⅝"	4"	17-19	12-14	150	6.0
5⅝"x7⅝"x11⅝"	6"	26-28	17-19	150	6.5
7⅝"x7⅝"x11⅝"	8"	33-35	21-23	150	7.0
9⅝"x7⅝"x11⅝"	10"	42-45	27-29	150	7.5
11⅝"x7⅝"x11⅝"	12"	48-51	29-31	150	8.0
3⅝"x3⅝"x15⅝"	4"	12-14	9-10	225	9.0
5⅝"x3⅝"x15⅝"	6"	17-19	11-13	225	9.5
7⅝"x3⅝"x15⅝"	8"	22-24	14-16	225	10.0
3⅝"x7⅝"x15⅝"	4"	23-25	16-18	112.5	5.0
5⅝"x7⅝"x15⅝"	6"	35-37	24-26	112.5	5.0
7⅝"x7⅝"x15⅝"	8"	45-47	29-31	112.5	6.0
9⅝"x7⅝"x15½"	10"	57-60	36-38	112.5	6.5
11⅝"x7⅝"x15⅝"	12"	64-67	40-42	112.5	7.0

Number of Concrete Building Units Laid Per 8-Hr. Day by One Mason

Actual Size of Units	Wall Thickness	Number Light-weight Units* Laid per 8-Hour Day	Hours Per 100 Pieces Mason	Hours Per 100 Pieces Labor
3⅝"x4⅞"x11⅝"	4"	215-235	3.5	3.5
5⅝"x4⅞"x11⅝"	6"	195-215	3.9	3.9
7⅝"x4⅞"x11⅝"	8"	175-195	4.3	4.3
3⅝"x7⅝"x11⅝"	4"	195-215	3.9	3.9
5⅝"x7⅝"x11⅝"	6"	175-195	4.3	4.3
7⅝"x7⅝"x11⅝"	8"	155-175	4.8	4.8
9⅝"x7⅝"x11⅝"	10"	135-155	5.5	5.5
11⅝"x7⅝"x11⅝"	12"	115-135	6.4	6.4
3⅝"x3⅝"x15⅝"	4"	215-235	3.5	3.5

Actual Size of Units	Wall Thickness	Number Lightweight Units* Laid per 8-Hour Day	Hours Per 100 Pieces Mason	Labor
5⅝"x3⅝"x15⅝"	6"	195-215	3.9	3.9
7⅝"x3⅝"x15⅝"	8"	175-195	4.3	4.3
3⅝"x7⅝"x15⅝"	4"	190-210	4.0	4.0
5⅝"x7⅝"x15⅝"	6"	170-190	4.5	4.5
7⅝"x7⅝"x15⅝"	8"	150-170	5.0	5.0
9⅝"x7⅝"x15⅝"	10"	130-150	5.7	5.7
11⅝"x7⅝"x15⅝"	12"	110-130	6.7	6.7
7⅝"x7⅝"x15⅝"	8"	225-250**	3.4	3.4
11⅝"x7⅝"x15⅝"	12"	120-160**	5.7	5.7

*For heavyweight concrete units decrease above quantities and increase labor 10 percent.
**For walls below grade.

INTERIOR STRUCTURAL CLAY FACING TILE

Facing tile discussed below includes those units at least one of whose faces, ceramic glazed, salt glazed or unglazed, are designed to be left exposed either as a partition, as backup to another type of facing, or as the inner wythe of a cavity wall.

These units are manufactured in accordance with sizes and specifications adopted by the Facing Tile Institute. The four sizes adopted are modular.

Facing tile are available in such a wide variety of colors and finishes as to be adaptable to most any design requirement, and the estimator must be knowledgeable about the product and grade and the desired end result.

Labor Laying Glazed Structural Facing Tile.—To insure the most economical construction, as well as the most attractive appearance in the finished work, all cutting of tile on the job should be done with a power saw using a carborundum blade.

The following labor quantities are based on the average job requiring not more than 20 percent of special shapes, such as bullnose corners, cove base, cap, etc. If the percentage of special pieces vary from the above, increase or decrease labor time proportionately.

Labor Laying Glazed Structural Facing Tile

Nominal Size of Tile*	Class of Work	No. Pcs. Set per 8-Hr. Day	Labor Hrs. per 100 Pcs. Mason	Labor
2"x5⅓"x 8"	Glazed one side	155-175	4.8	4.8
4"x5⅓"x 8"	Glazed one side	155-175	4.8	4.8
4"x5⅓"x 8"	Glazed two sides	125-155	5.7	5.7
6"x5⅓"x 8"	Glazed one side	155-175	4.8	4.8

6"x5⅓"x 8"	Glazed two sides	125-155	5.7	5.7
8"x5⅓"x 8"	Glazed one side...............	125-155	5.7	5.7
8"x5⅓"x 8"	Glazed two sides	105-120	7.0	7.0
2"x5⅓"x12"	Glazed one side...............	115-135	6.4	6.4
4"x5⅓"x12"	Glazed one side...............	115-135	6.4	6.4
4"x5⅓"x12"	Glazed two sides	90-110	8.0	8.0
6"x5⅓"x12"	Glazed one side...............	105-125	7.0	7.0
6"x5⅓"x12"	Glazed two sides	80-100	8.9	8.9
8"x5⅓"x12"	Glazed one side...............	85-100	8.7	8.7
8"x5⅓"x12"	Glazed two sides	65-85	10.6	10.6
2"x8 "x16"	Glazed one side...............	65-80	11.0	11.0
4"x8 "x16"	Glazed one side...............	55-65	13.3	13.3

Add for hoisting engineer if required.

*Nominal size includes thickness of standard mortar joint (¼") for glazed tile) in length, height and thickness.

Tile 1¾" thick is usually used for wall furring or partition wainscoting. Tile glazed one side is usually used for partitions that are glazed on one side and plastered on the other side. Tile glazed two sides are usually used for partition walls glazed on both sides of wall.

UNIT MASONRY/CERAMIC VENEER AND TERRA COTTA

Terra cotta is a building material made of a high grade clay (usually a mixture of several different kinds), to which is added a percentage of calcined clay to uniformly control the shrinkage and prevent undue warping.

Cost of Architectural Terra Cotta and Ceramic Veneer.—To be able to accurately arrive at the cost of these materials, an estimator must have an intimate knowledge of the manufacturing process. Each job is a custom job, there are no standard sizes, colors or shapes which are stocked.

When estimating quantities of terra cotta from the plans or details, all measurements should be squared; i.e., all molded courses, cornices, round columns, column caps, bases, coursed ashlar, sills, copings, etc., should be figured from the extreme dimensions and reduced to cubic feet.

The weight of terra cotta varies from about 80 lbs. per cu. ft. for plain 4" thick ashlar to 60 lbs. per cu ft. for large molded courses. For an average job about 72 lbs. per cu. ft., equivalent to 28 cu. ft. per ton of 2,000 lbs. may be used.

Cost of Handling and Setting Architectural Terra Cotta.—Inasmuch as a great deal of terra cotta consists of trimmings or 4" ashlar, it requires considerable handling to set a cubic foot. For instance, most ashlar being 4" thick, it is necessary to set terra cotta in 3 sq. ft. of wall to set one cu. ft. This also applies to sills, lintels, etc.

Based on the average job consisting of ashlar, sills, lintels, ornamental trimmings, etc., a mason will set about 35 to 40 cu. ft. (105 to 120 sq. ft. wall) per 8-hr. day, at the following labor cost per 100 cu. ft.:

	Hours	Rate	Total	Rate	Total
Terra cotta setter .	21	$....	$....	$17.87	$ 375.27
Helper......................	21	....	....	14.95	313.95
Labor sorting, handling wheeling, etc......	44	....	....	14.95	657.80
Cost per 100 cu. ft..........................			$....		$1,347.02
Cost per cu. ft.......				....	13.47
Cost per sq. ft (300)......................				....	4.49
Cost per ton (28 cu. ft.).................				....	377.17

On more complicated work consisting of gothic architecture, balusters and balustrade, glazed and enameled terra cotta, and other intricate work having a great deal of detail and requiring the utmost care in setting, a terra cotta setter will set only 20 to 25 cu. ft. (60 to 75 sq. ft. wall) per 8-hr. day, at the following labor cost per 100 cu. ft.:

	Hours	Rate	Total	Rate	Total
Terra cotta setter .	36	$....	$....	$17.87	$ 643.32
Helper......................	36	....	....	14.95	538.20
Labor sorting, handling, wheeling, etc......	66	....	....	14.95	986.70
Cost per 100 cu. ft..........................			$....		$2,168.22
Cost per cu. ft.......				....	21.68
Cost per sq. ft. (300)....................				....	7.23
Cost per ton (28 cu. ft.).................				....	607.10

Mortar Required for Setting Architectural Terra Cotta.—Under average conditions it will require about 4 cu. ft. of mortar to set one ton of terra cotta, or about ½-cu. yd. of mortar per 100 cu. ft.

GLASS BUILDING BLOCKS

Estimating Data on Laying Glass Blocks Using ¼-Inch Mortar Joints

	Size of Glass Blocks in Inches		
	5¾x5¾x3⅞	7¾x7¾x7⅞	11¾x11¾x3⅞
Cost per 100 Sq. Ft. of Wall			
Number of Blocks per Sq. Ft. of Wall*	4.00	2.25	1.00
Number of Blocks per 100 Sq. Ft. of Wall	400	225	100
Cu. Ft. of Mortar Required**	5	3.6	2.33
Mason Hours	21 to 23	14 to 16	10 to 12
Labor Hours	10½ to 11½	7 to 8	5 to 6
Labor on Scaffolding, Hours	2	2	2
Mason Ramming Oakum and Caulking, Hours	1½	1½	1½
Mason Cleaning Blocks, Hours	2½	2½	2½
Expansion Strips, ⅜"x4⅛", Lin. Ft.	30	30	30
Wall Ties, Lin. Ft.	60	60	60
Sponge Plastic Rope Joints, Lin. Ft.	460	60	60
Caulking Joints, Lin. Ft.	70	70	70
Asphalt Emulsion, 40'-0"x4½'	½ pt.	½ pt.	½ pt.
Cost per 1,000 Glass Blocks (Based on 10x10 Panels)			
Cu. Ft. of Mortar Required**	12.5	16	23.3
Mason Hours	53 to 57	62 to 71	100 to 120
Labor Hours	27 to 29	31 to 36	50 to 60
Labor on Scaffolding, Hours	5	9	20
Mason Ramming Oakum and Caulking, Hours	3¾	6¾	15
Mason Cleaning Blocks, Hours	6¼	11	25

Estimating Data on Laying Glass Blocks Using ¼-Inch Mortar Joints—Cont.

Cost per 100 Sq. Ft. of Wall	Size of Glass Blocks in Inches		
	5¾ x5¾ x3⅞	7¾ x7¾ x3⅞	11¾ x11¾ x3⅞
Expansion Strips, ⅜"x4⅛", Lin. Ft.	75	135	300
Wall Ties— Reinforcement, Lin. Ft.	110	200	440
Sponge Plastic Rope Joints, Lin. Ft.	150	266	600
Caulking Joints, Lin. Ft.	175	310	700
Asphalt Emulsion, 3½" Wide	1¼ pt.	2¼ pt.	2½ qts.
Number pcs. laid per 8-hr. day	140 to 150	115 to 125	70 to 80

*Does not include allowance for breakage.   **Includes 10 percent for waste.

04400 **STONE**

Labor Setting One Cu. Yd. of Rubble Stonework				
Description of Work	Cu. Ft. Stone	Cu. Ft. Mortar	Mason Hours	Labor Hours
Heavy rubble walls 2'-0" and over in thickness	27	7-9	3.0	3.0
Rubble walls up to 1'-6" in thickness	27	7-9	3.5	3.5

Labor Setting Random Ashlar in 100 Sq. Ft. of Wall	Sq. Ft. per 8-Hr. Day	Cu. Ft. Mortar	Mason Hours	Labor Hours
Setting random ashlar used as a veneer 4" thick, over wood framing and sheathing	32-35	10	24	24
Work above second story may require one extra laborer handling stone and getting it to setter				24
Setting random ashlar in straight work where stone requires backing and cutting end joints on job	17-25	10	38	38
Setting random ashlar in churches, chapels, etc., having numerous offsets, piers, pilasters, etc., where stone is backed in shop and requires very little job fitting	25-30	10	29	29
Setting random ashlar in churches, chapels, etc., where stone is sent to job with rough backs and requires backing and jointing on the job	14-18	10	50	50

Labor Setting Cut Stone

When estimating the labor cost of handling and setting cut stone, consider size of job, average size of stone, class of workmanship and method of setting, i.e., whether by hand, hand operated derricks or using stiff-leg, guy or other types of power derricks.

Description of Work	Cost per 100	Stone Setter	Helper Hours	Labor Hours
Setting Random or Range Ashlar in fine residences, churches, libraries, etc., consisting of 3" or 4" sawed strips, using an electric carborundum saw for jointing. 80 to 120 sq. ft. per 8-hr. day.	sq. ft.	8	8	8
Setting cut stone by hand on small jobs, consisting of sills, lintels, coping, steps, etc. 50 to 60 cu. ft. per 8-hr. day.	cu. ft.	14	....	28
Setting cut stone consisting of coursed ashlar, sills, lintels, etc., using hand operated breast derrick, gin pole, etc. 80 to 100 cu. ft. per 8-hr. day.	cu. ft.	9	9	36
Setting heavy cut stone consisting of heavy ashlar, molded courses, cornices, etc., using hand operated derricks. 115 to 140 cu. ft. per 8-hr. day.	cu. ft.	7	7	35
Setting heavy cut stone, consisting of coursed ashlar, sills, lintels, coping, etc. using a mobile power crane. 200 to 240 cu. ft. per 8-hr. day per setting crew hoisting engineer.	cu. ft.	4	8 / 4	20
Setting cut stone on large jobs using power operated stiff-leg or guy derricks, etc. 200 to 240 cu. ft. per 8-hr. day per setting crew. *Derrick and signal men. **Hoisting engineer.	cu. ft.	4	8* / 15*	25** / 5**

04500   MASONRY RESTORATION

Using ordinary methods, with an electric saw or ginder, a
mason will cut out about 25 sq. ft. of joints an hr. at the following labor cost per 100 sq. ft.:

	Hours	Rate	Total	Rate	Total
Mason	4	$....	$....	$18.91	$75.64
Cost per sq. ft			....		.76

After the mortar joints have been cut out, two tuckpointers
working on the scaffold, with one helper on the ground should
point 400 to 500 sq. ft. of brick wall per 8-hr. day, at the following labor cost per 100 sq. ft.:

	Hours	Rate	Total	Rate	Total
Tuckpointers	3.5	$....	$....	$18.91	$66.19
Helper	1.5	....	....	14.95	22.43
Cost per 100 sq. ft.			$....		$88.61
Cost per sq. ft			....		.89

SANDBLAST CLEANING OF BUILDINGS

The portable compressor and the sandblast are widely used
for cleaning buildings, bridges and other structures, either preparatory to repainting or for the purpose of brightening up the
surface.

With a portable compressor supplying air, the average operator can cover a strip 25'-0" wide and 60'-0" to 75'-0" high per 8-
hr. day. The exact area varies with the quality of the stone
encountered but the following table shows the speed at which
various stones can be cleaned.

	Sq. Ft. per Min.		Sq. Ft. per Min.
Limestone	7 to 9	Terra Cotta	8 to 10
Marble	3 to 5	Brick or Brownstone	8 to 10
Granite	6 to 8	Sand stone	10 to 12

Labor Cost of Sandblasting 100 Sq. Ft. of Old Brick or
Limestone Buildings.

	Hours	Rate	Total	Rate	Total
Compressor Operator	0.33	$....	$....	$17.95	$ 5.92
Nozzleman	0.33	....	....	13.04	4.30
Labor	0.67	....	....	12.54	8.40
Cost per 100 sq. ft.			$....		$18.62
Cost per sq. ft			....		.19

Add for Equipment rental, sand, and scaffold costs.

CHAPTER 5

METALS

CSI DIVISION 5

05100 STRUCTURAL METAL FRAMING

The structural steel framework of a building usually consists of anchor bolts, setting plates, base plates, columns, girders, beams, lintels, roof trusses, etc., which are fabricated from standard shapes, such as angles, "S" beams, channels, "W" beams and columns, plates, rods, etc., in combinations designed to give the required strength.

Estimating Quantities of Structural Steel

When estimating the quantity of structural steel required for any job, each class of work should be estimated separately, i.e., column bases, columns, girders, beams, lintels, trusses, etc., as they all involve different labor operations in fabrication and erection.

All standard connections can be taken from the A.I.S.C. Manual of Steel Construction. Be sure to include rivets and bolts for connecting steel to steel.

In addition to the structural framing of the building, it is often necessary for the structural steel contractor to furnish numerous miscellaneous items, such as bearing plates, loose lintels, anchor bolts, etc., although they are set in place by other contractors.

Erection of Structural Steel

Erecting Structural Steel in Skeleton Frame Buildings.—When figuring erection on a job of this type, consider carefully the amount of equipment, cranes, etc., required, as described previously.

On a one crane job, an erection gang should unload, handle, erect and connect 9 to 10 tons of steel per 8-hr. day, at the following labor cost per ton:

Erection gang	Hours	Rate	Total	Rate	Total
1 foreman	0.85	$....	$....	$18.20	$ 15.47
2 ir. wkrs.					
"hooking on"	1.70	....	....	17.70	30.09
1 ir. wkr. giving					
signals	0.85	....	....	17.70	15.05
4 ir. wkrs.					
"connecting"	3.40	....	....	17.70	60.18
Cost per ton			$....		$120.79

Add for workmen's compensation and liability insurance.

If the structural connections are welded instead of bolted, add $25.00 to $30.00 per ton.

ARC WELDING IN BUILDING CONSTRUCTION

Conditions Affecting Arc Welding Costs

Best results and lowest costs are obtainable only when proper plate preparation and welding procedure is followed. The following conditions have considerable bearing on the cost and quality of welded joints.

Fit-Up.—Care in cutting, forming and handling shapes to be welded, to avoid poor fit-up, is a major factor in the cost and performance of welded joints. However, a gap of 1/64" to 1/32" is useful in preventing angular distortion and weld cracking. The accompanying graph shows how various sizes of gaps affect welding speeds for square or grooved butt welds. Fillet welds are affected by oversize gaps in a similar manner.

Position of Joints.—The position of the joint has considerable effect on the speed and ease of welding. Wherever practical, welds should be made in the downhand position with the joint level. Vertical or overhead welds require much more time and skill.

Foreign Matter in Joint.—Excessive scale, paint, oil, rust, etc., all tend to interfere with welding and should be removed to obtain best speeds and results.

Build-Up or Overwelding.—Any amount of weld metal in addition to that actually needed for the specified strength is useless, costly, wasteful and, in some instances, actually harmful. For butt welds, there should be just enough build-up (no more than 1/16") to make sure weld is flush with the plate. Excessive build-up not only wastes weld metal but increases welding time.

Estimating the Weight of Wrought Iron, Steel, or Cast Iron

When tables of weights are not handy, the following rules will prove of value to the estimator when computing the weights of wrought iron, steel, or cast iron.

Weight of Wrought Iron.—One cubic foot of wrought iron weighs 480 pounds. One square foot of wrought iron 1-inch thick weighs 40 pounds. One square inch of wrought iron one foot long weighs 3 1/3 lbs.

To find the weight of one square foot of flat iron of any thickness, multiply the thickness in inches by 40, and the result will be the weight of the iron in pounds.

To find the weight of one lineal foot of wrought iron bar of any size, multiply the cross sectional area in square inches by 3 1/3, and the result will be the weight per lineal foot.

Weight of Steel.—One cubic foot of steel weighs 489.6 pounds, or 2 percent more than wrought iron. One square foot

WELDED JOINTS
Standard symbols

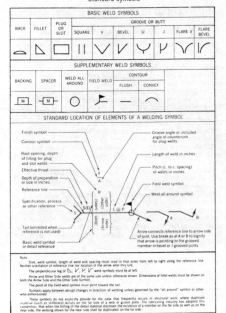

Courtesy of American Institute of Steel Construction, Inc.

of steel 1-inch thick weighs 40.8 pounds. A piece of steel 1-inch square and one foot long weighs 3.4 pounds.

To find the weight of one lineal foot of steel bar of any size, multiply the cross sectional area in square inches by 3.4, and the result will be the weight of the steel in pounds. If the weight per lineal foot is known, the exact sectional area in square inches may be obtained by dividing the weight by 3.4.

Weight of Cast Iron.—One cubic foot of cast iron weighs 450 pounds. One square foot of cast iron 1-inch thick weighs 37½ pounds. A piece of cast iron 1-inch square and one foot long weighs 3 1/8 pounds. One cubic inch of cast iron weighs .26 pound.

050200 METAL JOISTS

Standard open web and longspan steel joists are manufactured by welding chord members made up of hot rolled structural or cold formed sections to round bar or angle web members to form a truss. Information on the various types of steel joists can be obtained from the latest catalogs published by the Steel Joist Institute and the joist manufacturers.

Labor Setting Steel Joists.—The labor cost of handling and setting steel joists is usually estimated by the ton, based on the total weight of the joists to be placed.

This cost will vary considerably, depending upon the type of building, size of joists, length of spans, amount of handling and hoisting necessary, etc.

It is much cheaper to set joists on long straight spans and large floor areas, such as garage and factory buildings, office buildings, etc., than on small or other irregularly constructed buildings.

All steel joists must be bridged and connected according to the Steel Joist Institute standards.

On an ordinary job of steel floor joists of regular construction and fairly long spans, five men working together should handle and place about 4½ to 5 tons of joists (including bridging), per 8-hr. day, at the following labor cost per ton:

	Hours	Rate	Total	Rate	Total
Ironworker	40	$....	$....	$17.70	$708.00
Cost per ton				....	149.05

On jobs of irregular construction, five men working together should handle and place about 3½ to 4 tons of joists (including bridging), per 8-hr. day, at the following labor cost per ton:

	Hours	Rate	Total	Rate	Total
Ironworker	40	$....	$....	$17.70	$708.00
Cost per ton				....	188.80

Prices of Steel Joists.—Costs of standard steel joists will vary with several factors among which are the size of the job, the number of joists of one particular size and length, and the location of the job. For an average job the cost of standard joists will be in the range of $600.00 to $700.00 per ton, which would include nominal bridging members and other accessories.

Costs of longspan steel joists will vary with the same factors as given above for standard joists. For an average job the cost of the longspan joists will range from $580.00 to $750.00 per ton including accessories.

05300 METAL DECKING

Steel roof deck is formed from steel sheets in 18, 20, or 22 gauge. Most commercial decks have a standardized cross section with longitudinal ribs spaced 6" on centers. Sections are 30" in width and are manufactured in various lengths to suit job conditions, usually in the 14-ft. to 31-ft. range, although some can be supplied in greater lengths.

Deck sections are usually 1½" deep, but some manufacturers supply a 1¾", 2" or 2½" deep deck. Sections have interlocking or nesting side laps and telescoping or nesting end laps.

Under ordinary roof live loads—30 to 40 lbs. per sq. ft.—20 gauge deck is normally used for spans from 6'-0" to 7'-6" centers of purlins and 18 gauge deck for spans from 7'-0" to 8'-0". Specific information on allowable loads can be obtained from the various manufacturers. By extending a single sheet over two or more purlin spaces, structural continuity, increasing load carrying capacity, can be obtained. Dead load of deck, insulation and built-up roofing is approximately 7 lbs. per sq. ft.

Approximate cost of steel roof deck, delivered to job site in the eastern or northern states, on an average size job (10,000 sq. ft. or more) is as follows:

Gauge of Roof Deck	Price per Sq. Ft. 1½" Deck
22 Ga.	$ .70
20 Ga.	.80
18 Ga.	1.00

Placing Metal Deck on Top of Steel Joists.—When metal deck is placed over the top of steel joists to receive the concrete slab or lightweight insulating materials, the metal deck is usually spot welded to the joists.

On work of this type two men working together should place and weld about 200 sq. ft. of deck per hour at the following labor cost per square of 100 sq. ft.:

	Hours	Rate	Total	Rate	Total
Iron Worker	1.0	$....	$....	$17.70	$17.70
Cost per sq. ft.			....		.18

Reinforcing Floor Forms

Steel deck, erected in an inverted position (with ribs up) can be used as a reinforcing form for concrete construction. It acts as a form to support the concrete and permits the ribs to act as reinforcing.

Steel deck reinforcing forms can be erected for an entire structure as soon as the steel framework is placed, providing an immediate working platform and protective staging for all trades. Concrete can be placed at any portion of the building without regard for removing, cleaning and re-setting temporary formwork. Steel deck forms usually provide all necessary reinforcing to satisfy flexure requirements—the only additional reinforcing ordinarily required is temperature mesh to minimize shrink- age cracking. Tight form joints prevent concrete from dripping to lower floors, saving clean-up time.

Steel deck reinforcing forms can also be used in conjunction with composite beam design. Ample area for concrete around stud shear connectors and between deck ribs permits full effectiveness of the connectors. Standard A.I.S.C. Composite Design Procedure may be followed using the total slab depth in beam property calculations.

Reinforcing floor forms are installed by welding, the same as ordinary steel roof deck, and costs of materials and erection are about the same as given for "Metal Decking."

05400 LIGHTGAGE FRAMING

The components involved are a complete range of studs, joists and accessories for the steel framing of buildings. Sections are fabricated from structural-grade, high tensile strip steel by cold forming and are designed specifically for strength, light weight and low cost. Yet, structural framing carries all the benefits of conventional steel framing.

A nailing groove, developed for easy, economical attachment of other materials, is a feature of all double studs and joists. It is obtained by welding two cold-formed steel elements together, so designed that a nail driven into this space is not only held by friction, but is also deformed to provide maximum holding power.

The usual limitations imposed by prefabrication are avoided. Using nailable framing sections architects and engineers have unrestricted freedom in design. All sections are painted at the plant with a coat of oven-dried, rust-resisting red zinc chromate paint, or are galvanized.

Lightgage framing systems can supply complete wall, floor and roof construction for buildings up to four stories in height, or can be used in combination with other framing systems for interior, load bearing partitions, exterior curtain walls, fire separation walls, parapets, penthouses, trusses, suspended ceilings and mansard roofs.

The components can be completely detailed, cut and assembled in the shop and delivered for erection to the job site. Where on-the-job cutting and assembly is preferred, the material can be cut with a radial saw fitted with a $1/8$ " high speed circular blade.

Fastening may be by bolts, sheet metal screws or welding.

Joists come in 6", 8", 10", and 12" depths and in 12, 14, 16, and 18 gauge material. There are generally three styles: double nailable, "C" joists with 1⅝" flanges and "C" joists with 2½" flanges. Lengths from 6' to 40' are available. Joist webs may be solid, selectively punched or continuously punched for maximum raceway flexibility. Unnecessary punching should be avoided as the slotting lowers the structural value of the joist.

Joist bridging, which may be by stock "V" units or solid channels, must be supplied in the center of all spans up to 14'; at third points on spans from 14' to 20'; at quarter points on spans from 26' to 32'; and at eight foot centers on all spans over 32'.

Studs are available in 2½", 3⅝", 4" and 6" depths and in 14, 16, 18 and 20 gauges. 25 gauge studs are also available but are suitable for non-load bearing partitions only and are discussed under the drywall section of Chapter 9 "Finishes". Lengths from 7' to 40' are available with either slotted or solid webs. Studs may be of the double-nailable type, channels, or "C" type with knurled flanges.

05500   METAL FABRICATIONS

Metal Stairs.—Metal stairs are usually figured at so much per riser with stringers, treads, nosings, and railings included. A standard, three foot wide simple run metal pan stair can be budgeted at around $85.00 per riser, material cost. For each additional foot of width up to five feet add 10%. Custom stairs will run from $95.00 per riser to twice that amount where any of the stair components vary from the norm. Simple landings will run around $7.25 per square foot without an allowance for railings.

A four man crew should erect around 40 to 50 risers or 150 sq. ft. of landing per 8 hr. day.

The labor erection cost of a three foot wide steel stair consisting of three flights of sixteen risers and one 3' by 6' landing each run would figure as follows:

	Hours	Rate	Total	Rate	Total
Ironworkers -					
stairs 48 risers ...	34.0	$....	$....	$17.70	$601.80
landing 54 sq. ft.	11.1	....	....	17.70	196.47
Total Cost..............					$798.27
Cost per stair run (3).....................					266.09
Cost per riser incl. landings..............					16.63

Pipe Railings.—When estimating plain pipe railing, 3'-6" high, consisting of two horizontal runs of pipe with uprights 6 ft. to 8 ft. on center, figure as follows for the various sizes, erected in place:

1¼″ pipe railing, complete as described above, per lineal foot...$16.50

1½″ pipe railing, complete as described above, per lineal foot...$18.00

2″ pipe railing, complete as described above, per lineal foot...$20.00

For each curved section or termination point that is formed to radius, add $25.00 to $30.00.

For following the rake of a stair, add $7.50 per upright.

For each foot of wall railing $7.50

Aluminum pipe railings can also be ordered from stock and will cost around $30.00 per foot, anodized. Aluminum wall rails will cost around $8.50 per foot.

A two man crew should erect some 100 lineal feet of straight run railing per day, and 130 feet of wall railing.

Steel Ladders.—Straight steel ladders cost from $30.00 to $35.00 per lin. ft., erected in place. Add $30.00 to $40.00 for curved handles and platform over coping walls. Add $30.00 per lin. ft. if cage is required.

CHAPTER 6

CARPENTRY

CSI DIVISION 6

06100 ROUGH CARPENTRY

Estimating Wood Joists.—When estimating wood joists, always allow 4" to 6" on each end of the joist for bearing on the wall.

To obtain the number of joists required for any floor, take the length of the floor in feet, divide by the distance the joists are spaced and add 1 to allow for the extra joist required at end of span.

Example: If the floor is 28 ft. long and 15 ft. wide, it will require 16-ft. joists to allow for wall bearing at each end. Assuming the joists are spaced 16" on centers, one joist will be required every 16" or every 1 1/3 ft. In other words it will require 3/4 as many joists as the length of the span, plus one. Three-quarters of 28 equal 21, plus 1 extra joist at end, makes 22 joists 16 ft. long for this space.

The following table gives the number of joists required for any spacing:

Number of Wood Floor Joists Required for any Spacing

Distance Joists are Placed on Centers	Multiply Length of Floor Span by	Add Joists	Distance Joists are Placed on Centers	Multiply Length of Floor Span by	Add Joists
12 inches..	1	1	36 inches..	1/3 or .33	1
16 inches..	3/4 or .75	1	42 inches..	2/7 or .29	1
20 inches..	3/5 or .60	1	48 inches..	1/4 or .25	1
24 inches..	1/2 or .50	1	54 inches..	2/9 or .22	1
30 inches..	2/5 or .40	1	60 inches..	1/5 or .20	1

Estimating Quantity of Bridging.—It is customary to place a double row of bridging between joists about 6'-0" to 8'-0" on centers. Joists 10'-0" to 12'-0" long will require one double row of bridging or 2 pcs. to each joist.

Joists 14'-0" to 20'-0" long will require 2 double rows of bridging or 4 pcs. to each joist.

Bridging is usually cut from 1"x3", 1"x4", 2"x2", or 2"x4" lumber.

The following table gives the approximate number of pcs. and the lin. ft. of bridging required per 100 sq. ft. of floor.

Joists Up to 12 Feet Long				Joists Up to 20 Feet Long			
12" Centers		16" Centers		12" Centers		16" Centers	
No. Pcs.	Lin. Ft.	No. Pcs.	Lin. Ft.	No. Pcs.	Lin. Ft.	No. Pcs.	Lin. Ft.
20	30	16	24	40	60	32	48

Estimating Number of Wood Studs.—When estimating the number of wood partition studs, take the length of each partition and the total length of all partitions.

If a top and bottom plate is required, take the length of the wood partition and multiply by 2. The result will be the number of lin. ft. of plates required.

If a double plate consisting of 2 top members and a single bottom plate is used, multiply the length of the wood partitions by 3.

Number of Partition Studs Required for Any Spacing

Distance Apart Studs	Multiply Length of Partition by	Add Wood Studs
12 inches	1.0	1
16 inches	0.75	1
20 inches	0.60	1
24 inches	0.50	1

Add for top and bottom plates.

Quantity of Square Edged (S4S) Boards Required Per 100 Sq. Ft. of Surface

Measured Size Inches	Actual Size Inches	Add for Width	Ft. B.M. Req. per 100 Sq. Ft. Surface	Weight per 1000 Ft.
1x 4	$\frac{3}{4}$x 3$\frac{1}{2}$	14%	119	2300
1x 6	$\frac{3}{4}$x 5$\frac{1}{2}$	9%	114	2300
1x 8	$\frac{3}{4}$x 7$\frac{1}{4}$	10%	115	2300
1x10	$\frac{3}{4}$x 9$\frac{1}{4}$	8%	113	2300
1x12	$\frac{3}{4}$x11$\frac{1}{4}$	7%	112	2400

The above quantities include 5% for end cutting and waste.

Lineal Foot Table of Board Measure
Number of Feet of Lumber, B.M., Per Lineal Foot of any Size.

2" x 4"=0.667	4"x 4"=1.333	8"x14"= 9.333
2" x 6"=1.	4"x 6"=2.	8"x16"=10.667
2" x 8"=1.333	4"x 8"=2.667	10"x10"= 8.333
2" x10"=1.667	4"x10"=3.333	10"x12"=10.
2" x12"=2.	4"x12"=4.	10"x14"=11.667
2" x14"=2.333	4"x14"=4.667	10"x16"=13.333
2" x16"=2.667	4"x16"=5.333	10"x18"=15.
2$\frac{1}{2}$"x12"=2.5	6"x 6"=3.	12"x12"=12.
2$\frac{1}{2}$"x14"=2.917	6"x 8"=4.	12"x14"=14.
2$\frac{1}{2}$"x16"=3.333	6"x10"=5.	12"x16"=16.
3" x 6"=1.5	6"x12"=6.	12"x18"=18.
3" x 8"=2.	6"x14"=7.	14"x14"=16.333
3" x10"=2.5	6"x16"=8.	14"x16"=18.667
3" x12"=3.	8"x 8"=5.333	14"x18"=21.
3" x14"=3.5	8"x10"=6.667	16"x16"=21.333
3" x16"=4.	8"x12"=8.	16"x18"=24.

Lengths of Common, Hip, and Valley Rafters Per 12 Inches of Run

1	2	3	4	5*	6†
Pitch of Roof	Rise and Run or Cut	Length in Inches Common Rafter per 12" of Run	Percent Increase in Length of Com. Rafter over Run		Length in Inches Hip or Valley Rafters
1/12	2 & 12	12.165	.014	1.014	17.088
1/8	3 & 12	12.369	.031	1.031	17.233
1/6	4 & 12	12.649	.054	1.054	17.433
5/24	5 & 12	13.000	.083	1.083	17.692
1/4	6 & 12	13.417	.118	1.118	18.000
7/24	7 & 12	13.892	.158	1.158	18.358
1/3	8 & 12	14.422	.202	1.202	18.762
3/8	9 & 12	15.000	.250	1.250	19.209
5/12	10 & 12	15.620	.302	1.302	19.698
11/24	11 & 12	16.279	.357	1.357	20.224
1/2	12 & 12	16.971	.413	1.413	20.785
13/24	13 & 12	17.692	.474	1.474	21.378
7/12	14 & 12	18.439	.537	1.537	22.000
5/8	15 & 12	19.210	.601	1.601	22.649
2/3	16 & 12	20.000	.667	1.667	23.324
17/24	17 & 12	20.809	.734	1.734	24.021
3/4	18 & 12	21.633	.803	1.803	24.739
19/24	19 & 12	22.500	.875	1.875	25.475
5/6	20 & 12	23.375	.948	1.948	26.230
7/8	21 & 12	24.125	1.010	2.010	27.000
11/24	22 & 12	25.000	1.083	2.083	27.785
11/12	23 & 12	26.000	1.167	2.167	28.583
Full	24 & 12	26.875	1.240	2.240	29.394

*Use figures in this column to obtain area of roof surface for any pitch. Multiply number given above by footprint (plan) size of roof.

†Figures in last column are length of hip and valley rafters in inches for each 12 inches of common rafter run.

Table of Board Measure
Giving Contents in Feet of Joists, Scantlings and Timbers

Size in Inches	Length in Feet					
	10	12	14	16	18	20
1 x 2	12/3	2	21/3	22/3	3	31/3
1 x 3	21/2	3	31/2	4	41/2	5
1 x 4	31/3	4	42/3	51/3	6	62/3
1 x 6	5	6	7	8	9	10
1 x 8	62/3	8	91/3	102/3	12	131/3
1 x10	81/3	10	112/3	131/3	15	162/3

	10	12	14	16	18	20
1 x12	10	12	14	16	18	20
11/4x 4	41/6	5	55/6	62/3	71/2	81/3
11/4x 6	61/4	71/2	83/4	10	111/4	121/2
11/4x 8	81/3	10	112/3	131/3	15	162/3
11/4x10	105/12	121/2	147/12	162/3	183/4	205/6
11/4x12	121/2	15	171/2	20	221/2	25
11/2x 4	5	6	7	8	9	10
11/2x 6	71/2	9	101/2	12	131/2	15
11/2x 8	10	12	14	16	18	20
11/2x10	121/2	15	171/2	20	221/2	25
11/2x12	15	18	21	24	27	30
2 x 2	31/3	4	42/3	51/3	6	62/3
2 x 3	5	6	7	8	9	10
2 x 4	62/3	8	91/3	102/3	12	131/3
2 x 6	10	12	14	16	18	20
2 x 8	131/3	16	182/3	211/3	24	262/3
2 x10	162/3	20	231/3	262/3	30	331/3
2 x12	20	24	28	32	36	40
2 x14	231/3	28	322/3	371/3	42	462/3
3 x 4	10	12	14	16	18	20
3 x 6	15	18	21	24	27	30
3 x 8	20	24	28	32	36	40
3 x10	25	30	35	40	45	50
3 x12	30	36	42	48	54	60
3 x14	35	42	49	56	63	70
4 x 4	13	16	19	21	24	27
4 x 6	20	24	28	32	36	40
4 x 8	27	32	37	43	48	53
4 x10	33	40	47	53	60	67
4 x12	40	48	56	64	72	80
4 x14	47	56	65	75	84	93
6 x 6	30	36	42	48	54	60
6 x 8	40	48	56	64	72	80
6 x10	50	60	70	80	90	100
6 x12	60	72	84	96	108	120
6 x14	70	84	98	112	126	140
6 x16	80	96	112	128	144	160
8 x 8	53	64	75	85	96	107
8 x10	67	80	93	107	120	133
8 x12	80	96	112	128	144	160
8 x14	93	112	131	149	168	187
8 x16	107	128	149	171	192	213
10 x10	83	100	117	133	150	167
10 x12	100	120	140	160	180	200
10 x14	117	140	163	187	210	233
10 x16	133	160	187	218	240	267
12 x12	120	144	168	192	216	240
12 x14	140	168	196	224	252	280
12 x16	160	192	224	256	288	320

Table of Board Measure-Cont.
Giving Contents in Feet of Joists, Scantlings and Timbers

Size in Inches	Length in Feet					
	10	12	14	16	18	20
14 x14		163	196	229	261	294	327
14 x16		187	224	261	299	336	373
14 x18		210	252	294	336	378	420
14 x20		233	280	327	373	420	467
16 x16		213	256	299	341	384	427
16 x18		240	288	336	384	432	480
16 x20		267	320	373	425	480	533
18 x18		270	324	378	432	486	540
18 x20		300	360	420	480	540	600
20 x20		333	400	467	533	600	667

Nails Required for Carpenter Work

The following table gives the number of wire nails in pounds for the various kinds of lumber per 1,000 ft., board measure, or per 1,000 shingles and lath or per square (100 sq. ft.) of asphalt slate surfaced shingle, with the number of nails added for loss of material on account of lap or matching of shiplap, flooring, ceiling and siding of the various widths. The table gives the sizes generally used for certain purposes with the nailing space 16" on centers, and 1 or 2 nails per board for each nailing space.

Description of Material	Unit of Measure	Size and Kind of Nail	Number of Nails Required	Pounds of Nails Required
Wood Shingles	1,000	3d Common	2,560	4 lbs.
Individual Asphalt Shingles	100 sq. ft.	7/8" Roofing	848	4 lbs.
Three in One Asphalt Shingles	100 sq. ft.	7/8" Roofing	320	1 lb.
Wood Lath	1,000'	3d Fine	4,000	6 lbs.
Wood Lath	1,000'	2d Fine	4,000	4 lbs.
Bevel or Lap Siding, 1/2"x4"	1,000'	6d Coated	2,250	*15 lbs.
Bevel or Lap Siding, 1/2"x6"	1,000'	6d Coated	1,500	*10 lbs.
Byrkit Lath, 1"x6"	1,000'	6d Common	2,400	15 lbs.

Drop Siding, 1"x6"..........	1,000'	8d Common	3,000	25 lbs.
3/8" Hardwood Flooring.......	1,000'	4d Common	9,300	16 lbs.
25/32" Hardwood Flooring.......	1,000'	8d Casing	9,300	64 lbs.
Subflooring, 1"x3"...........	1,000'	8d Casing	3,350	23 lbs.
Subflooring, 1"x4"...........	1,000'	8d Casing	2,500	17 lbs.
Subflooring, 1"x6"...........	1,000'	8d Casing	2,600	18 lbs.
Ceiling, 5/8"x4"........	1,000'	6d Casing	2,250	10 lbs.
Sheathing Boards, 1"x4"..........	1,000'	8d Common	4,500	40 lbs.
Sheathing Boards, 1"x6"..........	1,000'	8d Common	3,000	25 lbs.
Sheathing Boards, 1"x8"..........	1,000'	8d Common	2,250	20 lbs.
Sheathing Boards, 1"x10".........	1,000'	8d Common	1,800	15 lbs.
Sheathing Boards, 1"x12".........	1,000'	8d Common	1,500	12 1/2 lbs.
Studding, 2"x4"............	1,000'	16d Common	500	10 lbs.
Joist, 2"x6".....	1,000'	16d Common	332	7 lbs.
Joist, 2"x8".....	1,000'	16d Common	252	5 lbs.
Joist, 2"x10"...	1,000'	16d Common	200	4 lbs.
Joist, 2"x12"...	1,000'	16d Common	168	3 1/2 lbs.
Interior Trim, 5/8" thick....	1,000'	6d Finish	2,250	7 lbs.
Interior Trim, 3/4" thick....	1,000'	8d Finish	3,000	14 lbs.
5/8" Trim where nailed to jamb........	1,000'	4d Finish	2,250	3 lbs.
1"x2" Furring or Bridging..	1,000'	6d Common	2,400	15 lbs.
1"x1" Grounds	1,000'	6d Common	4,800	30 lbs.

---

*NOTE—Cement coated nails sold as two-thirds of pound equals 1 pound of common nails.

Recommended Nailing Schedule for Common Applications in Building Construction

Application	Nailed Into	Nail Size Inches	Nail Type	Head Diameter	Head Type	Point Size Type / Dia.	Nail-ing	Spacing o.c.	Nails Per Joint
Mudsill, partition plate, 2"	Concrete	21/2-23/4x0.148	Sc-1z	5/16"	Checkered	Long Dia.	Face	12"-24"	—
Ditto, in earthquake regions	Concrete	(31/4-) 31/2x0.250	Sc-1z	9/16"	Checkered	Med. Ndl.	Face	24"-48"	—
Ditto, 3"	Concrete	41/2x0.250	Sc-1z	9/16"	Checkered	Med. Ndl.	Face	24"-48"	—
Furring strips	Concrete	11/2-13/4x0.148	Sc-1z	5/16"	Checkered	Long Dia.	Face	12"-24"	—
Mudsill	Mudsill	21/2x0.120	Sc-2	9/32"	Flat	Med. Dia.	Toe	—	2
Sleepers	Mudsill	21/2x0.120	Sc-2	9/32"	Flat	Med. Dia.	Toe	—	2
Joists	Mudsill	31/4x0.135	Sc-2	5/16"	Flat	Med. Dia.	Toe	—	2-3
Subflrg., 1" lumber, plywood	Mudsill, sleeper, joist	21/8x0.105	St-14	1/4"	Flat, Csk.	Med. Dia.	Face	6" & 12"	2(3)
Subflrg. 2" lumber, plywood	Mudsill, sleeper, joist	27/8x0.120	St-14	9/32"	Flat, Csk.	Med. Dia.	Face	—	2(3)
Subflrg, 3/8"-1/2" plywood (dph.)	Mudsill, sleeper, joist	11/2x0.135	Hi-28	5/16"	Flat, Csk.	Med. Dia.	Face	6" & 12"	—
Subflrg. 5/8" plywood (dph.)	Mudsill, sleeper, joist	13/4x0.135	Hi-28	5/16"	Flat, Csk.	Med. Dia.	Face	6" & 12"	—

Subflrg, 3/4" plywood (dph.), part bd.	Mudsill, sleeper, joist	2 x0.148	Hi-28	5/16" Flat, Csk.	Med. Dia.	Face	6" & 12"	—	—
Subflrg, 1"-11/8" plywood (dph.), part bd	Mudsill, sleeper, joist	21/2x0.148	Hi-28	5/16" Flat, Csk.	Med. Dia.	Face	6" & 12"	—	—
Underlayment, 1/4"-5/16fm)" plywood	Subfloor	1 x0.083	St-16	3/16" Flat, Csk.	Med. Dia.	Face	6" & 6"-12"	—	—
Underlayment, 3/8"-1/2" plywood	Subfloor	11/4x0.083	St-16	3/16" Flat, Csk.	Med. Dia.	Face	6" & 6"-12"	—	—
Underlayment, 5/8" plywood	Subfloor	13/8x0.098	St-16	1/4" Flat, Csk.	Med. Dia.	Face	6" & 6"-12"	—	—
Underlayment, 3/4" plywood	Subfloor	11/2x0.098	St-16	1/4" Flat, Csk.	Med. Dia.	Face	6" & 6"-12"	—	—
Underlayment, 7/8" plywood	Subfloor	15/8x0.098	St-16	1/4" Flat, Csk.	Med. Dia.	Face	6" & 6"-12"	—	—
Underlayment, 3/16"-5/8" hardboard	Subfloor	1-13/8x0.083	St-15	3/16" Flat, Csk.	Med. Dia.	Face	6" & 12"	—	—
Flooring, T & G hardwood	Subfloor, joist, sleeper	2-21/2x0.115	Sc-4	13/64" Casing	Blunt Dia.	Toe	10"-18"	—	—
Flooring, T & G softwood	Subfloor, joist, sleeper	2-21/2x0.115	Sc-4	13/64" Casing	Blunt Dia.	Toe	10"-18"	—	—
Flooring, T & G hardwood, 3/8" and 1/2"	Subfloor, joist, sleeper	1-11/4x0.072	Sc-4	9/64" Casing	Blunt Dia.	Toe	10"-18"	—	—
Flooring, T & G parquet	Subfloor	11/2x0.105	Sc-4	9/64" Casing	Blunt Dia.	Face	—	—	—
Framing plates	Stud	31/4x0.135	Sc-2	5/16" Flat	Med. Dia.	Face	—	—	2

Recommended Nailing Schedule for Common Applications in Building Construction-Cont.

Application	Nailed Into	Nail Size Inches	Nail Type	Head Diameter	Head Type	Point Size Dia. Type	Nail-ing Type	Spacing o.c.	Nails Per Joint
Framing studs	Stud, cripple, lintel, sill	2½"x0.120	Sc-2	9/32"	Flat	Med. Dia.	Face	16"-24"	—
Framing studs	Plate, cripple, lintel, sill	2½"x0.120	Sc-2	9/32"	Flat	Med. Dia.	Toe	—	3
Framing sole plate	Mudsill	3¼"x0.135	Sc-2	5/16"	Flat	Med. Dia.	Face	16"	—
Framing top plate	Lower top plate	3¼"x0.135	Sc-2	5/16"	Flat	Med. Dia.	Face	24"	—
Trussed rafter assembly		3¼"x0.135	Sc-5	5/16"	Flat	Med. Dia.	Face	2½"-3"	Given
Trussed rafter assembly		2½"x0.120	Sc-5	9/32"	Flat	Med. Dia.	Face	2½"-3"	Given

See Notes, Key to Nail Types and Abbreviations on later page.

Application	Nailed Into	Nail Size Inches	Nail Type	Head Diameter	Head Type	Point Size Dia. Type	Nail-ing Type	Spacing o.c.	Nails Per Joint
Rafter, 4"	Top plate	3¼"x0.135	Sc-2	5/16"	Flat	Med. Dia.	Toe	—	3
Rafter, 4"	Top plate	6 x0.177	St-34	7/16"	Flat	Med. Dia.	Face	—	2
Rafter, 4"	Top plate	7 x0.207	St-34	1/2"	Flat	Med. Dia.	Face	—	2
Rafter, 6", 8", 10"	Top plate	4-6 x0.177 / 7-9 x0.203	St-34	7/16"	Flat	Med. Dia.	Toe	—	2-3
Sheathing, 1" lumber	Framing, rafter	2 x0.120	Sc-3	9/32"	Flat	Med. Dia.	Face	—	2
Sheathing, 3/8"-1/2" plywood	Framing, rafter	1¾x0.120	St-17	9/32"	Flat	Med. Dia.	Face	6" & 12"	—
Sheathing, 5/16"-1/2" plywood (dph.)	Framing, rafter	1½x0.135	Hi-17	5/16" Flat, Csk.		Med. Dia.	Face	6" & 12"	—

Sheathing, 5/8" plywood (dph.)	Framing, rafter	13/4x0.135 Hi-17	5/16" Flat, Csk.	Med. Dia.	Face	6" & 12"	—
Sheathing, 3/4" plywood (dph.)	Framing, rafter	2 x0.148 Hi-17	5/16" Flat, Csk.	Med. Dia.	Face	6" & 12"	—
Sheathing, 1" 11/8" plywood (dph.)	Framing, rafter	21/2x0.148 Hi-17	5/16" Flat, Csk.	Med. Dia.	Face	6" & 12"	—
Sheathing, insulation board, gypsumboard	Framing, rafter	11/2-2x0.120 St-10g	3/8", 7/16" Flat	Blunt Dia.	Face	3-4" & 6-8"	—
Sheathing, asbestosboard, 1/8"	Framing, rafter	11/4x0.063 St-8g	3/16" Flat, Csk.	Blunt Dia.	Face	3-4" & 6-8"	—
Sheathing, asbestosboard, 1/4"	Framing, rafter	11/4x0.120 St-or Sc-6g	5/16" Flat, Csk.	Blunt Dia.	Face	3-4" & 6-8"	—
Sheathing, hardboard, 3/8"-5/8"	Framing, rafter	2 x0.115 Sc-7g	13/64", Flat, Csk.	Med. Ndl.	Face	3-4" & 6-8"	—
Building paper	Sheathing	1/2-3/4x0.105 Sq-30	15/16" Square	Med. Dia.	Face	6"-12"	—
Stripping, 3/8"x35/8"	Framing, joist, rafter	2 x0.120 St-3	9/32" Flat	Med. Dia.	Face	—	2
Stripping, 1"x4"	Framing, joist, rafter	21/2x0.135 St-3	5/16" Flat	Med. Dia.	Face	—	2
Stripping, 2"x3"	Framing, joist, rafter	31/2x0.165 St-3	5/8" Flat	Med. Dia.	Face	—	2
Siding, wood, 1"	Sheathing and framing	21/8x0.101-0.115 St-14g	1/4" Flat, Csk.	Med. Dia.	Face	—	—
Siding, wood, 1"	Sheathing and framing	2 x0.120 Dr-14	5/32" Flat, Csk.	Med. Ndl.	Face	—	1

Recommended Nailing Schedule for Common Applications in Building Construction-Cont.

Application	Nailed Into	Nail Size Inches	Nail Type	Head Diameter Type	Point Size Type	Nailing	Spacing o.c.	Nails Per Joint
Siding, wood, 2"	Sheathing and framing	3 x0.135	Dr-14	5/32" Flat, Csk.	Med. Ndl.	Face	—	1
Siding, plywood	Sheathing and framing	17/8x0.109	Dr-8	5/32" Casing	Med. Ndl.	Face	6" & 12"	1
Siding, T & G wood	Sheathing and framing	13/4x0.105	Sc-8	5/32" Casing	Med. Dia.	Toe	6" & 12"	—
Siding, asbestos shingle	Sheathing and framing	11/2-13/4x0.105	Dr-33	3/16" Flat Button	Med. Dia.	Face	—	Given
Siding, asbestos shingle	Sheathing and framing	11/2-13/4x0.083	St-19t	3/16" Flat	Med. Dia.	Face	—	Given
Siding, asbestos shingle	Sheathing and framing	11/2-13/4x0.076	St-20	3/16" Flat	Med. Dia.	Face	—	Given
Siding, insulated brick, wood shingle	Sheathing and framing	13/4x0.095	St-18ge	3/16" Flat	Med. Dia.	Face	8"-12"	2
Siding, wood shingle	Insulating sheathing	13/4-2 x0.083	St-18ge	5/32" Finishing	Blunt Dia.	Face	—	2
Siding, wood shingle	Insulating sheathing	13/4-2 x0.105	Dr-18	5/32" Finishing	Blunt Dia.	Face	—	2
Siding, wood shingle	Plywood	11/8x0.102	Dr-18	3/16" Flat	Long Dia.	Face	—	2

Siding, hardboard	Framing	2-21/2x0.115 Sc-7z	13/64"	Casing	Long Ndl.	Face	12"	—	
Siding, hardboard battenboard	Framing	11/2x0.083 Sc-7z	9/64"	Casing	Long Ndl.	Face	12"	—	

See Notes, Key to Nail Types and Abbreviations on later page.

Fascia, 1"	Framing, rafter	21/2x0.120 Sc-2g	9/32"	Flat	Med. Dia.	Face	12"	—	2
Fascia, 2" lumber	Framing, rafter	31/4x0.135 Sc-2g	5/16"	Flat	Med. Dia.	Face	12"	—	2
Roofing, built-up	Sheathing	3/4-11/4x0.105 Sq-30	15/16"	Square	Med. Dia.	Face	10"	—	
	Poured gypsum	11/2-13/4x0.120 Sq-31	15/16"	Square	Med. Dia.	Face	10"	—	
Roofing, built-up	Sheathing	3/4-2x0.120 St-10g	3/8"	Flat	Blunt Dia.	Face			2-3
Roofing, asphalt shingle	Sheathing	3/4-2x.120-.135 Dr-10	3/8"	Flat	Blunt Dia.	Face			2-3
Roofing, asphalt shingle	Sheathing	3/4-2x.105-.120 Dr-18	3/16"	Flat	Blunt Dia.	Face			2
Roofing, wood shingle	Sheathing	13/4-2x0.083 St-18g	1/8"	Flat, Csk.	Blunt Dia.	Face			2
Roofing, wood shingle	Sheathing	As for siding							
Roofing, asbestos shingle	Rafter, purlin	11/2-13/4x0.145 Dr-10	13/32"	Flat	Long Dia.	Face	12"	—	
Roofing, aluminum (corr. and flat)		1-3x0.135 St-or	7/16"	Flat	Long Dia.	Face	12"	—	
Roofing, sheet metal (corr. and flat)	Rafter, purlin	11/2-3x0.135 Sc-9g	7/16"	Flat	Long Dia.	Face	12"	—	
Roofing, glass fiber (corr. and flat)	Rafter, purlin	11/2-3x0.148 Dr-9 or 10	7/16"	Flat	Long Dia.	Face	12"	—	*With Neoprene washer attached*

Recommended Nailing Schedule for Common Applications in Building Construction-Cont.

Application	Nailed Into	Nail Size Inches	Nail Type	Head Diameter	Head Type	Point Size	Point Type	Nail-ing	Spacing o.c.	Nails Per Joint
Lath, expanded metal, K-lath	Framing, joist	11/2x0.148	St-22g		L-Shaped		Med. Dia.	Face	6" & 12"	—
Lath, gypsum plasterboard	Framing, joist	11/4x0.101	St-23b	19/64"	Flat, Csk.		Long Dia.	Face	5"	—
Gypsumboard, 3/8"	Framing, joist	11/4x0.098	St-24	1/4"-19/64"	Flat, Csk.		Long Dia.	Face	5"-8"	—
Gypsumboard, 1/2"-5/8"	Framing, joist	13/8x0.098	St-24	1/4"-19/64"	Flat, Csk.		Long Dia.	Face	5"-8"	—
Gypsumboard, prefinished	Framing, joist	13/8x0.083	K-32e	3/16" Flat,	Csk.		Long Dia.	Face	5"-8"	—
Paneling, trim		1-11/4x0.054	K-32e	3/32"			Blunt Dia.	Face	—	—
Paneling, trim, exterior		1-11/2x0.072	St-13	3/32"	Casing		Blunt Dia.	Face	—	—
Paneling, trim, exterior		1 x0.065	St-12	3/32"	Casing		Blunt Dia.	Face	—	—
Paneling, trim, exterior		11/2x0.076	Sc-12	3/32" Oval	Casing		Blunt Dia.	Face	—	—

Paneling, trim		1 x0.072	Sc-11		3/32" Casing	Blunt Dia.	Face	—	—
Paneling, trim		11/2 13/4x0.083	Sc-11		1/8" Casing	Blunt Dia.	Face	—	—
Paneling, trim		21/2x0.105	Sc-11		9/64" Casing	Blunt Dia.	Face	—	—
Acoustic tile		1-13/4x0.062	St-25z		—	Blunt Dia.	Face	—	—
Electric conduit	Wood	11/2-2 x0.162	St-26z		1" Hook	Blunt Ndl.	Face	—	—
Electric conduit	Masonry	11/2-2 x0.162	St-27z		1" Hook	Blunt Ndl.	Face	—	—
Fencing wire	Softwood (treated)	11/2x0.148	St-22g		L-Shaped	Med. Dia.	Face	—	—
Fencing wire	Hardwood	11/2x0.148	St-21g		L-Shaped	Med. Dia.	Face	—	—

See Notes, Key to Nail Types and Abbreviations on next page.

Key to Nail Types

Sc-iz	Screw-Tite Masonry Nail, hardened HCS, zinc plated	St-17	Stronghold Sheathing Nail, bright LCS
Sc-2	Screw-Tite Framing Nail, hardened HCS	St-18g	Stronghold Shingle Nail, bright LCS, galvanized
Sc-2g	Screw-Tite Framing Nail, hardened HCS, galvanized	St-18ge	Stronghold Shingle Nail, bright LCS, galvanized and enameled
Sc-3	Screw-Tite Framing Nail, bright LCS	St-19t	Stronghold Shingle Nail, bronze, tin plated
Sc-3g	Screw-Tite Framing Nail, bright LCS, galvanized	St-20	Stronghold Shingle Nail, stainless steel
Sc-4	Screw-Tite Flooring Nail, hardened HCS	St-21g	Stronghold Fence Staple, hardened HCS, galvanized
Sc-5	Screw-Tite Trussed Rafter Nail, hardened HCS	St-22g	Stronghold Fence Staple, bright LCS, galvanized
Sc-6g	Screw-Tite Asbestosboard Nail, hardened HCS, galvanized	St-22z	Stronghold Fence Staple, bright LCS, zinc plated
Sc-7g	Screw-Tite Exterior Hardboard Nail, hardened HCS, galvanized	St-23b	Stronghold Lath Nail, bright LCS, blued
Sc-7z	Screw-Tite Exterior Hardboard Nail, hardened HCS, zinc plated	St-24	Stronghold Drywall Nail, bright LCS
Sc-8	Screw-Tite Casing Nail, silver bronze	St-25z	Stronghold Kollarnail, hardened HCS, zinc plated
Sc-9g	Screw-Tite Roofing Nail, hardened HCS, galvanized	St-26z	Stronghold Conduit Staple, bright LCS, zinc plated
Sc-10g	Screw-Tite Roofing Nail, bright LCS, galvanized	St-27z	Stronghold Knurled Conduit Staple, hardened HCS, zinc plated

Sc-11	Screw-Tite Finishing Nail, bright LCS
Sc-12	Screw-Tite Finishing Nail, stainless steel
St-3	Stronghold Framing Nail, bright LCS
St-4	Stronghold Parquet Flooring Nail, hardened HCS
St-6g	Stronghold Asbestosboard Nail, hardened HCS, galvanized
St-9g	Stronghold Roofing Nail, hardened HCS, galvanized
St-10g	Stronghold Roofing Nail, bright LCS, galvanized
St-12	Stronghold Finishing Nail, stainless steel
St-13	Stronghold Finishing Nail, monel metal
St-14	Stronghold Sinker Nail, bright LCS
St-14g	Stronghold Sinker Nail, bright LCS, galvanized
St-15	Stronghold Underlay Nail, Hardened HCS
St-16	Stronghold Underlay Nail, bright LCS
St-34	Stronghold Spike, hardened HCS
Hi-28	"Hi-Load" Shear-Resistant Nail, bright LCS
Hi-17	"Hi-Load" Sheathing Nail, bright LCS
Sq-30	Squared Annular Thread Cap Nail, bright LCS
Sq-31	Squared Spiral Thread Cap Nail, bright LCS
K-32e	Annular Thread Kolorpin, bright LCS, enameled
Dr-8	Drive-Rite Spiral Thread Casing Nail, aluminum
Dr-9	Drive-Rite Screw Thread Roofing Nail, aluminum
Dr-10	Drive-Rite Spiral Thread Roofing Nail, aluminum
Dr-14	Drive-Rite Spiral Thread Sinker Nail, aluminum
Dr-18	Drive-Rite Screw Thread Shingle Nail, aluminum
Dr-20	Drive-Rite Spiral Thread Shingle Nail, aluminum
Dr-33	Drive-Rite Knurled Asbestos-Cement Shingle Face, aluminum

NOTES: For fastening redwood, use only aluminum or stainless steel nails.

Local conditions, customs and popular usage may dictate minor variations in length and gauge of nails. Consult the Technical Service Department of Independent Nail & Packing Company, Bridgewater, Mass.

The above chart is based on a table appearing in Bulletin No. 38 (Revised Edition). "Better Utilization of Wood Through Assembly with Improved Fasteners," a study undertaken at Wood Research Laboratory, Virginia Polytechnic Institute, under the sponsorship of Independent Nail & Packing Company, Bridgewater, Mass., manufacturers of Stronghold Annular Thread and Screw-Tite Spiral Thread Nails and other improved fasteners.

STRONGHOLD ® ANNULAR THREAD NAIL

STRONGHOLD ® SCREW THREAD NAIL

SCREW-TITE ® SPIRAL THREAD NAIL

SCREW-TITE ® KNURLED MASONRY NAIL

ABBREVIATIONS USED IN THIS TABLE:
Cort.—Corrugated
Dph.—Diaphragm
Part. bd.—Particle board
LCS—Low Carbon Steel
HCS—High Carbon Steel
Plywd.—Plywood
Csk.—Countersunk   Med.—Medium
Flrg.—Flooring   Dia.—Diamond
Shgl.—Shingle   Ndl.—Needle

Nails Required for Subflooring.—On wood floors up to 4" wide, quantities are based on 8d flooring nails. For flooring 6" and wider, 10d nails have been figured.

The quantities given below are sufficient to lay 1,000 ft. of flooring, b.m.

Width Flooring	Joist Spacing 12" on Centers	Joist Spacing 16" on Centers
2"	40 lbs. 8d flg.	30 lbs. 8d flg.
3"	30 lbs. 8d flg.	23 lbs. 8d flg.
4"	22 lbs. 8d flg.	17 lbs. 8d flg.
6"	24 lbs. 10d com.	18 lbs. 10d com.
8"	17 lbs. 10d com.	13 lbs. 10d com.

Data on Common Wire Nails

Size of Nails	Length of Nails Inches	Gauge Number	Approximate Number to Pound	Approx. Price Per 100 lbs.
4d	1½	12½	316	$36.00
5d	1¾	12½	271	36.00
6d	2	11½	181	36.00
8d	2½	10¼	106	34.00
10d	3	9	69	34.00
12d	3¼	9	63	34.00
16d	3½	8	49	34.00
20d	4	6	31	34.00
30d	4½	5	24	34.00
40d	5	4	18	34.00
50d	5½	3	14	34.00
60d	6	2	11	34.00

Check local market as these prices are subject to rapid change.

Hardware Accessories Used for Wood Framing

Steel Joist Hangers.—Used for framing joists to beams and around openings for stair wells, chimneys, hearths, ducts, etc. Made of galv. steel, varying from 12 ga. to 3/16" in thickness, with square supporting arms, holes punched for nails and with bearing surfaces proportioned to size of lumber. Approximate prices for sizes most commonly used are as follows:

Joist Size	Gauge	Depth of seat	Opening in Hanger	Price Each
2"x 6"	12	2"	15/8"x5"	$0.70
2"x 8"	12	2"	15/8"x5"	.78
2"x10"	12	2"	15/8"x81/2"	.81
2"x12"	10	21/2"	15/8"x81/2"	1.00
4"x 8"	11	2"	35/8"x51/4"	.94
4"x10"	9	2"	35/8"x51/4"	1.12
4"x12"	3/16	21/2"	35/8"x81/2"	1.73

Teco* Framing Accessories.—Used in light wood construction to provide face-nailed connections for framing members. Adaptable to most framing connections, they eliminate the uncertainties and weaknesses of toe-nailing. Manufactured of zinc-coated, sheet steel in various gauges and styles. Framing accessories are designed to provide nailing on various surfaces. Special nails, approximately equal to 8d common nails, but only 1¼" long, to prevent complete penetration of standard nominal 2" lumber, are furnished with anchors. Approximate prices are as follows:

Type	Price Per 100	Type	Price Per 100
Trip-L-Grips	$25.00	Post Caps	$66.00
Du-Al-Clip	20.00	H Clips	4.00
Truss Plates	30.00	Angles	45.00

WOOD ROOF TRUSSES

Wood roof trusses are used in buildings where clear floor space is a requirement and the width of the building exceeds the economical span of roof joists. Some of the advantages of trussed roof construction are fast erection, unobstructed weather protected space available sooner, simplified installation of ceiling, floor, mechanical and electrical systems, use of non-load-bearing movable partitions, etc.

Wood roof trusses are used for spans as short as 25'-0", and may be used up to 200'-0".

When using wood roof trusses, it is common practice to span the shorter dimension of the building. This may be accomplished with any of several truss designs available, such as bowstring, crescent, Belgian, flattop, super-bow, scissors and Gothic. The type of truss is dependent upon the degree in which economy is emphasized, the use of the building and the general architectural effect desired.

Prices on roof trusses are governed by the following conditions:

1. Cost of material and labor.
2. Loading conditions and spacing.
3. Difficulties of erection.
4. Requirements of local building ordinances.

Spacing of wood roof trusses is usually 16'-0" to 20'-0" on centers, but the 16'-0" spacing is more generally used. The spacing of trussed rafters which support the roof sheathing directly (instead of with purlins as is the case where trusses are used) is usually 2'-0". Where roof loads are light and the installation of a ceiling is not required, spacings of 4'-0" and 5'-0" can be advantageous. Where snow loads are especially heavy, spacings of 16" and even 12" have been used.

Labor Erecting Wood Roof Trusses.

Under average working conditions on jobs requiring 5 or more trusses, a crew consisting of 5 men should erect one truss in the following time:

Length of Truss	No. Crew	Labor Hrs. per Truss	Hours	Rate	Total	Rate	Total
50'-0"	1	5	5	$....	$....	$16.47	$ 82.35
60'-0"	1	5	5	....	....	16.47	82.35
75'-0"	2	10	10	....	....	16.47	164.70
90'-0"	3	15	15	....	....	16.47	247.05
100'-0"	4	20	20	....	....	16.47	329.40
125'-0"	5	25	25	....	....	16.47	411.75

WOOD BLOCKING, FURRING AND GROUNDS

Placing Wood Furring Strips on Masonry Walls.—Where it is necessary to place wood furring strips on brick or tile walls before lathing and plastering, allowing the strips to follow the line of the walls without wedging or blocking out to make them straight or plumb, with the nails driven into dry joints in the brickwork, a carpenter should place 500 to 550 lin. ft. of furring per 8-hr. day, at the following labor cost per 100 lin. ft. :

	Hours	Rate	Total	Rate	Total
Carpenter	1.5	$....	$....	$16.47	$24.71
Cost per lin. ft.					.25
Cost per Square (100 Sq. Ft.)					
Strips 12" centers.	1.7	$....	$....	$16.47	$27.99
Strips 16" centers.	1.3	....	....	16.47	21.41

Placing Wood Floor Sleepers.—When placing 2"x3" or 2"x4" wood floor screeds or sleepers over rough tile or concrete floors, to receive finish flooring, a carpenter should place 225 to 275 lin. ft. per 8-hr. day, at the following labor cost per 100 lin. ft. :

	Hours	Rate	Total	Rate	Total
Carpenter	3.2	$....	$....	$16.47	$52.70
Labor	0.8	....	....	12.54	10.03
Cost 100 lin. ft.			$....		$62.73
Cost per lin. ft.			....		.63

BUILDING AND INSULATING SHEATHING

Labor Placing Insulating Sheathing.—When placing insulating sheathing on square or rectangular houses of regular

Labor Framing and Erecting Lumber In Rough Carpentry

Description of Work	Ordinary Work		First Grade Work	
	Carp. Hours	Labor Hours	Carp. Hours	Labor Hours
Framing and placing foundation wall plates, per 100 lin. ft.	6.4	......	......	......
Framing and setting wood sills and plates, 2"x4" and 2"x6"—100 lin. ft.	6.4	......	8.0	8.0
Framing and erecting exterior stud walls for plain square or rectangular buildings per 1000 b.m.	21.4	6.0	32.0	8.0
Framing and erecting exterior stud walls for irregularly shaped residences, such as English type, etc.	29.1	6.0	45.7	8.0
Framing and erecting interior stud partitions*	20.0	6.0	29.1	8.0
Framing and setting floor joists up to 2"x8"**	13.9	4.5	15.2	4.5
Framing and setting floor joists 2"x10" or 2"x12"**	12.8	4.0	13.9	4.0
Framing panel and girder floor systems*	10.0	5.0		
Installing cross-bridging, per 100 sets	8.0			
Framing and erecting rafters on plain gable roofs*	24.6	7.0		
Framing and erecting rafters for gable roofs cut up with dormers, etc.*	29.1	8.0		
Framing light timbers for exposed roof beam construction*	20.0	10.0		

*Per 1000 ft. b.m.

Labor Framing and Erecting Lumber In Rough Carpentry

Description of Work	Ordinary Work Carp. Hours	Labor Hours
Framing and erecting rafters for plain hip roofs*	29.1	8.0
Framing and erecting rafters for difficult hip roofs, containing dormers, gables, valleys, etc.*	40.0	8.0
Framing light timbers for exposed roof beam construction*	20.0	10.0
Placing wood cant strips, per 100 lin. ft.	1.6	
Framing roof saddles for flat roofs	19.0	6.0
Laying rough wood floors of 1"x6" or 1"x8"*	9.2	5.3
Laying rough wood floors 1"x6" or 1"x8" laid diagonally*	11.5	5.5
Laying roof sheathing on plain hip or gable roofs*	13.5	6.5
Laying sheathing on steep roofs and roofs cut up with dormers, hips, valleys, etc.*	26.5	6.5
Placing 1"x6" or 1"x8" sidewall sheathing*	12.0	5.0
Placing 1"x6" or 1"x8" sidewall sheathing diagonally*	16.5	5.5
Laying Wood Sheathing on Flat Roofs	9.2	5.3
Laying Plywood Subflooring Per 100 Sq. Ft.	0.9	0.3
2"x4' wood blocking, if reqd. per 100 sq. ft.	1.8	0.5
Laying Plywood Decking for Panel and Girder Floor Systems per 1000 sq. ft.	1.75	3.50
Laying Wood Sheathing on Pitch or Gable Roofs*	13.5	6.5
Placing Sidewall Sheathing*	16.5	5.5
*per 1000 ft. b.m.		

construction, a carpenter should place 700 to 900 sq. ft. per 8-hr. day, at the following labor cost per 100 sq. ft. :

	Hour	Rate	Total	Rate	Total
Carpenter	1.0	$....	$....	$16.47	$16.47
Labor unloading and					
carrying sheets ..	0.3	....	....	12.54	3.76
Cost per 100 sq. ft			$....		$20.23
Cost per sq. ft			....		.20

PLYWOOD ROOF AND WALL SHEATHING, SUBFLOORING AND UNDERLAYMENT

Roof and Wall Sheathing—Plywood for use in roof and wall sheathing is readily available, easily worked with ordinary tools and skills, and is adaptable to almost any light construction application. Because of the large size sheets, installation time is less than for other types of sheathing, and waste is minimal. It may be used under any type of shingle or roofing material, or any type of siding.

Standard size sheets are 4"x8" but other sizes are available by special order. Most common sheathing thicknesses are 5/16", 3/8", 1/2", 5/8" and 3/4", which are unsanded, and are available with interior or exterior glue.

For most roof and wall sheathing installations, where sheathing is to be covered, interior type plywood is used. Where installations are to be exposed to the weather, such as for roof overhangs or service building siding, exterior type plywood should be used.

Nailing of plywood sheathing should be at 6" o.c. along panel edges and 12" o.c. at intermediate supports. 6d common nails should be used for panels ½" or less in thickness, and 8d for greater thickness.

Where plywood is used for wall sheathing, corner bracing is not required because of the rigidity of the plywood. Where other types of sheathing are used, plywood of the same thickness as the sheathing is sometimes used at corners to take the place of corner diagonal bracing as shown in the photograph.

Subflooring and Underlayment.—Nailing of subflooring should be at 6" o.c. along the edges and 10" o.c. at intermediate supports. 6d common nails may be used for ½" plywood, and 8d for ⅜" to ⅞" thick plywood. For 1⅛" thick panels, use 8d ring shank or 10d common nails spaced 6" o.c. at both panel edges and intermediate supports.

Underlayment grades of plywood are touch sanded panels of ¼" and ⅜" thickness with a smooth, solid surface for application of nonstructural flooring finished directly to them. This grade or plywood is made with special inner-ply construction which resists punch-through by concentrated point loading.

Combined subfloor-underlayment plywood—is a single application installation combining both subfloor and underlayment, thus eliminating the expense of installing a separate underlayment. Panel edges perpendicular to the joists must be supported by solid blocking if tongue and grooved panels are not used.

Panels should be nailed at 6" o.c. at edges and 10" at intermediate supports with ringshank or spiral-thread nails. 6d deformed-shank nails may be used for panel thickness up to ¾" and 8d for thicker panels.

Approximate costs of plywood sheathing, subflooring and underlayment per 1000 sq. ft. are as follows:

Interior Type Plywood STANDARD (C-D INT)		Interior Type Plywood Structural 1 C-D		Exterior Type Plywood C-C EXT	
5/16"	$240.00	1/2"	$310.00	5/16"	$290.00
3/8	260.00	5/8	390.00	3/8	350.00
1/2	310.00	3/4	470.00	1/2	390.00
5/8	360.00			5/8	430.00
3/4	420.00			3/4	500.00

Plywood Underlayment Group 1, Interior

1/2"	$300.00
5/8	390.00
3/4	450.00

Gypsum Sheathing.—Gypsum sheathing is a fireproof solid sheet of gypsum encased in a tough, fibrous, water-resisting covering. The sides and ends are treated to resist moisture. Some manufacturers blend asphalt emulsion into the wet gypsum core mix to provide additional resistance to moisture. It is used for sheathing on frame structures under siding, shingles, stucco and brick veneer.

Gypsum sheathing has V-joint edges on the long dimension to provide positive assurance of a tight fit at the unsupported joints. It is applied with its length at right angles to the studs.

Gypsum sheathing is ½-in. thick, 2 ft. wide and 8 and 10 ft. long, and 4 ft. wide and 8 and 9 ft. long to fit supports 16 in. on centers. Nails should be 1¾-in. long, No. 10½ gauge, galvanized flat head roofing nails, spaced 4-in. on centers, except under wood siding and stucco, 8". Requires 14 to 21 lbs. per 1,000 sq. ft.

Framing and Erecting Lumber in Heavy Mill Construction

Description of Work	Carp. Hours	Labor Hours
Time Given per 1,000 Ft. of Lumber, b. m.		
Framing and erecting heavy wood columns or posts to 10"x10"–10'-0" to 16'-0" long	35.0	8.0
Same as above, labor hours per column	3.7	0.5
If hoisted above second floor, add per column		0.5
Rounding corners on large timbers, per 1,000' b. m. $15.00 to $20.00		
Boring holes through center of large timbers, $25.00 to $30.00 per 1,000'		
Framing and erecting heavy wood beams for first floor	7.0	7.0
If necessary to frame for post caps, bases and stirrups, add	14.5	7.0
Framing and erecting heavy wood beams and girders above 1st. floor	16.0	10.0
Framing and erecting heavy wood floor joists, 4"x12" to 4"x16"	7.3	3.5
Framing and erecting heavy wood floor joists, 3"x10" to 4"x10"	9.0	4.0
Framing and Placing Oak Bumpers at Edges of Loading Docks, 4"x6" to 12"x12"	48.0	24.0
Laying 2"x6" or 3"x6" Tongue and Groove Timber Subfloors or Roof Sheathing	8.9	6.0
Laying 2"x6" or 3"x6" D&M wood sub-floors on large areas	7.3	5.0
Laying 2"x6" or 3"x6" D&M wood sub-flooring in small spaces requiring cutting and fitting		
Laying 2"x4" or 2"x6" laminated wood floors	9.5	5.0
For irregular spans requiring considerable cutting, add 20 to 25%	9.0	5.0
Laying 2"x8" to 2"x12" laminated wood floors	8.0	5.0
For irregular spans requiring considerable cutting, add 20 to 25%		

Labor Cost of 100 Sq. Ft. of Gypsum Sidewall Sheathing
Applied to Square or Rectangular Buildings of Regular
Construction

	Hours	Rate	Total	Rate	Total
Carpenter sheathing	1.0	$....	$....	$16.47	$16.47
Helper unloading and carrying sheets	0.3	....	....	12.54	3.76
Carpenter applying paper	0.4	....	....	16.47	6.59
Cost per 100 sq. ft.			$....		$26.82
Cost per sq. ft.			....		.27

06200 FINISH CARPENTRY

Placing Corner Boards, Fascia Boards, Etc.—When placing wood fascia boards, corner boards, etc., on houses, cottages, etc., a carpenter should place 175 to 225 lin. ft. per 8-hr. day, at the following labor cost per 100 lin. ft.:

	Hours	Rate	Total	Rate	Total
Carpenter	4	$....	$....	$16.47	$65.88
Cost per lin. ft.			....		.66

Placing Exterior Wood Cornices, Verge Boards, Etc.—When placing exterior wood cornices, verge boards, fascia, soffits, etc., consisting of 2 members, two carpenters working together should place 150 to 175 lin. ft. per 8-hr. day, at the following labor cost per 100 lin ft.:

	Hours	Rate	Total	Rate	Total
Carpenter	10	$....	$....	$16.47	$164.70
Cost per lin. ft.			....		1.65

Framing and Erecting Exterior Wood Stairs for Rear Porches.—When framing and erecting outside wood stairs for rear porches on apartment buildings, etc., where the stringers are 2"x10" or 2"x12" lumber, with treads and risers nailed on the face of the stringers, it will require 18 to 22 hrs. carpenter time per flight of stairs.

This is for an ordinary stair having 14 to 18 risers extending from story to story, and the labor per flight should cost as follows:

	Hours	Rate	Total	Rate	Total
Carpenter	20	$....	$....	$16.47	$329.40

If the stair consists of 2 short flights with an intermediate landing platform between stories it will require 12 to 13 hrs. carpenter time per flight or 24 to 26 hrs. per story, including platform, at the following labor cost:

	Hours	Rate	Total	Rate	Total
Carpenter	25	$....	$....	$16.47	$411.75

Placing Plywood Soffits.—Two carpenters should set 75 sq. ft. of ¼" and 64 sq. ft. of ½" plywood on soffits per hour at the following labor cost per 100 sq. ft.:

	Hours	Rate	Total	Rate	Total
Carpenter	2.8	$....	$....	$16.47	$45.92
Cost per sq. ft ...			....		.46

Placing Wood Base.—This cost will vary with the size of the rooms and whether a single member or 2 or 3-member base is specified.

Where there are 55 to 60 lin. ft. of 2 member base in each room without an unusually large amount of cutting and fitting, a carpenter should place 125 to 150 lin. ft. per 8-hr day, at the following labor cost per 100 lin. ft.:

	Hours	Rate	Total	Rate	Total
Carpenter	5.8	$....	$....	$16.47	$ 95.53
Labor	1.0	....	....	12.54	12.54
Cost per 100 lin. ft.			$....		$108.07
Cost per lin. ft			....		1.08

If there are an unusually large number of miters to make, such as required in closets and other small rooms, increase the above costs accordingly.

Placing Wood Chair or Dado Rail.—In large rooms, long, straight corridors, etc., a carpenter should fit and place 275 to 300 lin. ft. of wood chair rail per 8-hr. day, at the following labor cost per 100 lin. ft.:

	Hours	Rate	Total	Rate	Total
Carpenter	2.8	$....	$....	$16.47	$46.12
Labor	0.5	....	....	12.54	6.27
Cost per 100 lin. ft.			$....		$52.39
Cost per lin. ft			....		.52

Placing Wood Cornices.—Where 3 or 4-member wood cornices are placed in living rooms, reception rooms, dining rooms, etc., a carpenter should place cornice in one average

sized room per 8-hr. day, which is equivalent to 50 to 60 lin. ft.,
and the labor cost per 100 lin. ft. would be as follows:

	Hours	Rate	Total	Rate	Total
Carpenter	14.5	$....	$....	$16.47	$238.82
Labor	2.0	....	....	12.54	25.08
Cost per 100 lin. ft			$....		$263.90
Cost per lin. ft					2.64

Setting Cabinets and Cases.—When setting mill assembled
cabinets and cases, such as used in the average residence or
apartment building, it will require about 1/6 hr. carpenter time
for each sq. ft. of cabinet face area at the following labor cost
per sq. ft. :

	Hours	Rate	Total	Rate	Total
Carpenter	0.16	$....	$....	$16.47	$2.64

**Setting Factory Assembled and Finished Kitchen Cabi-
nets.**—When setting factory assembled and finished kitchen
cabinets, such as used in the average apartment or residence, it
will require about 0.1-hr. carpenter time for each sq. ft. of cabinet face area at the following labor cost per sq. ft. :

	Hours	Rate	Total	Rate	Total
Carpenter	0.1	$....	$....	$16.47	$1.65

Setting Wood Fireplace Mantels.—If wood fireplace mantels are factory assembled, merely requiring fitting and setting,
a carpenter should install 2 to 3 mantels per 8-hr. day at the
following labor cost per mantel :

	Hours	Rate	Total	Rate	Total
Carpenter	3.2	$....	$....	$16.47	$52.70
Labor	1.0	....	....	12.54	12.54
Cost per mantel		....	$....		$65.24

Fitting and Placing Closet Shelving.—A carpenter should
fit and place 115 to 135 sq. ft. of closet shelving per 8-hr. day
including setting of shelf cleats at the following labor cost per
100 sq. ft. :

	Hours	Rate	Total	Rate	Total
Carpenter	6.4	$....	$....	$16.47	$105.41
Cost per sq. ft					1.05

Installing Hanging Rods in Closets.—A carpenter should
install about three hanging rods, including supports, per hr. at
the following labor cost per rod :

	Hours	Rate	Total	Rate	Total
Carpenter	0.33	$....	$....	$16.47	$5.44

Labor Placing Finish Hardware

Description of Work	No. Set per 8-Hr. Day	Carp. Hrs. Each
Rim locks or night latches, cheap work	20-24	0.3-0.4
Mortised locks in soft or hardwood doors, Ordinary Workmanship	12-16	0.5-0.7
Mortised locks in hardwood doors, First Grade Workmanship	6-8	1.0-1.3*
Cylindrical locks	17-23	0.3-0.5
Front entrance cylinder locks in hardware doors, Ordinary Workmanship	6-8	1.0-1.3**
Front entrance cylinder locks in hardwood doors, First Grade Workmanship	4-5	1.6-2.0***
Panic bolts, First Grade Workmanship	2-3	2.7-4.0
Door closers, exposed	7-9	0.9-1.1
Door closers, concealed	2-3	2.7-4.0
Door holders	20-24	0.3-0.4
Sash lifts and locks, No. of windows	20-24	0.3-0.4
Cremone bolts for casement doors and windows	20-24	0.3-0.4
Kickplates, Ordinary Workmanship	8-10	0.8-1.0
Kickplates, First Grade Workmanship	6-8	1.0-1.3

*On this grade of work the lock must be flush with the face of the door and must operate perfectly.

**On this class of work the architect or owner must expect to have more or less trouble with hardware because the time given is not sufficient to adjust lock and see that it operates properly, etc.

***This includes front door locks having fancy escutcheon plates, knobs, knockers, etc.

06300 WOOD TREATMENT

Wood and plywood are often given special treatments at the mill to give additional protection against fire, rot and termites.

Fire protected or fire retardant wood is pressure-impregnated with mineral salts that will react chemically at a temperature below the ignition point of the wood. An Underwriter's Laboratory Flame Spread Rating of 25 can be obtained, and no periodic maintenance is required to keep the rating. However the lumber must also be kiln dried to 12% moisture content if it is to be painted. Also, wood is not suitable for direct exposure to weather and ground conditions unless further treatment is given. A fire retardant treatment will add 12 cts. to 15 cts. per board foot to lumber costs. If it is necessary to kiln dry the lumber add 6 cts for soft woods, 8 cts. for hard woods.

Rot and termite resistance can be obtained by several methods. Creosote, a coal tar product, is impregnated at the rate of around 8# per cubic foot at a cost of 12 cts. to 15 cts. per board foot. This finish is not readily paintable and can stain adjacent finishes such as plaster and wallboard. Pentachlorophenol is a popular oil bourne treatment, and costs 11 cts. to 13 cts. per sq. ft. and leaves the wood with an oil residue. This residue can be eliminated and the wood left clean and paintable if the pantachlorophenal is impregnated into the wood by liquid petroleum. This will add 5 cts. to the cost.

INTERIOR PANELING

Placing Sheet Paneling.—The actual cost of placing sheet paneling will vary with the size and shape of the room, whether full size sheets may be used or considerable cutting and fitting necessary. It requires practically as much time to place a sheet 16"x96" as one 48"x96" on account of the cutting and fitting, while there are three times as many sq. ft. in the latter as in the former.

On straight work, in large rooms, a carpenter should fit and place 450 to 550 sq. ft. per 8-hr. day, at the following labor cost per 100 sq. ft. :

	Hours	Rate	Total	Rate	Total
Carpenter	1.6	$....	$....	$16.47	$26.35
Cost per sq. ft			....		.26

When applying sheets in smaller rooms requiring considerable cutting and fitting around doors, windows, etc., a carpenter should fit and place 275 to 300 sq. ft. per 8-hr. day, at the following labor cost per 100 sq. ft. :

	Hours	Rate	Total	Rate	Total
Carpenter	2.7	$....	$....	$16.47	$44.47
Cost per lin. ft			....		.44

06423 Solid Wood Paneling

Solid paneling is usually quoted in random widths of 4" to 8" and random lengths of 6' to 16'. Specified widths and lengths will run 10% or more additional. Traditional thickness is ¾", but much paneling today is available at a considerable savings in ½" thickness.

There is no nationally established grading rules for solid wall paneling, so it is necessary to know the product of the mill quoting to compare prices. Generally, woods are available in two grades: "select," "luxury" or "clear," and "knotty," "wormy" or "pecky," often designated as "Colonial" grade.

In figuring paneling, add 15% to actual areas for waste and matching.

Paneling is available with V-joint tongue and groove, square joint, shiplap and Colonial beaded.

Placing Wood Wainscot or Paneling (Knocked Down).—If wood wainscot or paneling is sent to the job knocked down, making it necessary to assemble the pieces and put them together on the job, a carpenter should place top and bottom rails, cap and intermediate panels for 10 to 15 lin. ft. of wainscoting per 8-hr. day, at the following costs for the different heights:

Height of Wainscot	No. Sq. Ft. 8-Hr. Day	Carp. Hrs. 100 Sq. Ft.
4'-0"	50-60	13-16
5'-0"	60-70	11-13
6'-0"	70-80	10-11
7'-0"	80-90	9-10
8'-0"	90-100	8- 9

CHAPTER 7

THERMAL AND MOISTURE PROTECTION

CSI   DIVISION 7

07100   WATERPROOFING

Some water and dampness resisting preparations are furnished in mastic form, and are applied with a trowel. Others are mixed with portland cement and sand and applied as a plaster coat to the surfaces to be treated.

There are also numerous water and dampness resisting compounds in the form of heavy paints, which are applied to the concrete or masonry walls, floors, and ceilings with a brush or mop. One or more applications of these materials are supposed to penetrate and fill the pores in the concrete or masonry surfaces to such an extent that the treated surfaces will be impervious to dampness or moisture.

WATER AND DAMP-RESISTING PLASTER COATS

There are numerous preparations on the market to be mixed with portland cement and sand and applied as a plaster coat for waterproofing concrete and masonry walls. There are other preparations in mastic form that are applied with a trowel, the same as ordinary cement mortar.

Quantity of Cement and Sand Required for One Cu. Yd. of Cement Mortar

Based on damp loose sand containing 5 percent of moisture and weighing 2,565 lbs. per cu. yd.

Proportions by Volume Cement Sand	Packed Cement Sacks	Loose Sand Cu. Yds.	Proportions by Volume Cement Sand	Packed Cement Sacks	Loose Sand Cu. Yds.
1 :1	18.0	0.70	1 :2½	11.0	1.03
1 :1½	15.2	0.84	1 :3	9.5	1.10
1 :2	12.5	0.95			

Above quantities of sand include 5 percent for waste.

Number of Sq. Ft. of Cement Plaster Coat Obtainable from One Cu. Yd. of Cement Mortar

Thickness in Inches	Square Feet	Thickness in Inches	Square Feet
¼	1,296	¾	432
½	648	1	324

Labor Applying Waterproof Plaster Coats to Vertical Surfaces.

A man should apply about 350 sq. ft. of first coat per 8-hr. day, at the following labor cost per 100 sq. ft. :

	Hours	Rate	Total	Rate	Total
Plasterer	2.3	$....	$....	$16.35	$37.61
Helper	2.3	....	....	13.77	31.67
Cost per 100 sq. ft.			$....		$69.29
Cost per sq. ft.					.69

On the second or finish coat, a man should apply about 400 sq. ft. per 8-hr. day, at the following labor cost per 100 sq. ft.:

	Hours	Rate	Total	Rate	Total
Plasterer	2.0	$....	$....	$16.35	$32.70
Helper	2.0	....	....	13.77	27.54
Cost per 100 sq. ft.			$....		$60.24
Cost per sq. ft.			....		.60

If necessary to clean the old masonry or concrete surfaces, or to roughen the old concrete, add for this work.

Water Resisting Plaster Coats Using Powdered Intregal Water Repellents, Per 100 Sq. Ft.

Trade Name	Inches Thick	Mix Prop.	Number Coats	Sack Cement	Cu. Ft. Sand	Quantity Waterp'g	Mechanic Hours	Labor Hours
Hydratite	1	1 :2	2	4	8	8 lbs.	8	8
Hydrocide Powder	1	1 :2	2	4	8	4 lbs.	8	8
Toxement IW	¾	1 :2	2	3	6	4½ lbs.	7	7

Water Resisting Plaster Coats Using Liquid Intregal Admixtures, Per 100 Sq. Ft.

Trade Name	Inches Thick	Mix Prop.	Number Coats	Sack Cement	Cu. Ft. Sand	Quantity Waterp'g	Mechanic Hours	Labor Hours
Anti-Hydro	¾	1 :2	2	3	6	¾ gal.	7	7
Anti-Hydro	1	1 :2	2	4	8	1 gal.	8	8
Aquatite	¾	1 :2	2	3	6	¾ gal.	7	7
Aquatite	1	1 :2	2	4	8	1 gal.	8	8
Hydratite	¾	1 :2	2	3	6	¾ gal.	7	7
Hydratite	1	1 :2	2	4	8	1 gal.	8	8
Toxement IW	¾	1 :2	2	3	6	¾ gal.	7	7
Toxement IW	1	1 :2	2	4	8	1 gal.	8	8
Sikacrete	¾	1 :2	2	3	6	¾ gal.	7	7
Trimix	¾	1 :2	2	3	6	¾ gal.	7	7

MEMBRANE WATERPROOFING

Membrane waterproofing is constructed in place by building up a strong, waterproof and impermeable blanket with overlapping plies of tar or asphalt saturated open mesh cotton fabric or rag felt. The plies are coated and cemented together with hot coal tar pitch or waterproofing asphalt. There is always one or more applications of pitch or asphalt than plies of felt or fabric.

Estimating the Quantity of Felt or Fabric Required for Membrane Waterproofing.

Fabric or Felt Required to Cover 100 Sq. Ft. of Surface

Class of Work	1 Ply 4" lap	1 Ply 6" lap	2-Ply	3-Ply	4-Ply	5-Ply	Ea. Add. Ply
Add percentage Sq. Ft.	20	25	10	10	10	10	10
Felt	120	125	220	330	440	550	110

Estimating the Quantity of Pitch or Asphalt Required for Membrane Waterproofing.

Weight of Pitch or Asphalt Per 100 Sq. Ft. of Surface

No. Plies of Saturated Fabric or Tarred Felt	Alternate Moppings of Pitch or Asphalt Required	Lbs. of Pitch or Asphalt Required 100 Sq. Ft. Waterproofing
2	3	90 to 105
3	4	120 to 140
4	5	150 to 175
5	6	180 to 210
6	7	210 to 245

Prior to the application of the first hot bitumen mopping coat, the wall surface is primed, uniformly and completely, with one gallon of primer (asphalt primer for asphalt specifications or tarbase primer for pitch specifications) per 100 sq. ft. of surface area.

For the application of the primer ot the surface, one man should coat about 100 sq. ft. per hour, at the following labor cost:

	Hours	Rate	Total	Rate	Total
Labor	1.0	$....	$....	$16.46	$16.46
Cost per sq. ft.					.17

Labor Applying Membrane Waterproofing.—When applying membrane waterproofing to walls and floors, three or four men usually work together, as it requires one man to attend the fire and heat the bitumen, while two or three men are mopping the walls and applying the felt or fabric.

Labor Applying 100 Sq. Ft. of Membrane Waterproofing

Number Plies	Description of Work	Mopping Hours	Felt Hours	Total Hours
1	1-ply fabric, 2 moppings	1.6	0.8	2.4
2	2-ply fabric, 3 moppings	2.4	1.6	4.0
3	3-ply fabric, 4 moppings	3.2	2.4	5.6
4	4-ply fabric, 5 moppings	4.0	3.2	7.2
1	Each additional ply of felt and mopping	0.8	0.8	1.6

THE IRON METHOD OF WATERPROOFING

The iron method of waterproofing uses an exceedingly fine metallic powder, containing no grease, asphalt, oil or other substances subject to disintegration.

The waterproofing is applied to the inside or outside surfaces of walls and to the tops of rough footings and floor slabs, in the form of brush coats or in a combination of brush and plaster coats.

Applied either upon the inside or outside of walls, this type of waterproofing will resist hydrostatic pressure to a considerable degree.

When preparing old concrete vertical surfaces for iron waterproofing, the vertical surfaces should be given an entirely new bonding exposure by cutting not less than 1/16" with bush-hammers or other suitable tools. See cost of roughing old surfaces on previous pages. The new surface shall then be thoroughly cleaned by brushing with wire brushes. Surfaces of brick and stone need only thorough cleaning with the wire brush and washing.

Holes, cracks, and soft or porous spots in vertical or horizontal surfaces shall be cut out and pointed with one part iron floor bond, 2 parts portland cement and 3 parts sand, all by weight. These materials shall be mixed dry, screened and enough water added to make a stiff mix. Particular care must be taken at the intersections of all walls and floors, in corners, around pipes and other projections through the waterproofing and at construction joints in the concrete.

After the wall surfaces have been cleaned and pointed, they shall be thoroughly cleaned again by hosing. Excess water shall then be removed.

It usually requires 2 to 6 coats of iron waterproofing, depending upon the density of the surface and the conditions encountered.

Some of these materials are mixed with sand and water and applied as a brush coat while others are furnished already mixed ready to apply by the addition of water.

1-Coat Waterproofing Using the
Iron Method, Brushed On, Per 100 Sq. Ft.

Trade Name	Number Coats	Quan. Req'd 100 Sq. Ft.	Approx. Cost Lb.	Labor Hrs. Applying
Ferrocon (Castle)....	1	10–15 lbs.	$0.13	2
Ferrolith W..............	1	13–15 lbs.	.25	2
Kemox H (Sika)........	1	60 lbs.–100 lbs.	.14	2
Irontox.....	1	10–12 lbs.	.19	2

For each additional application, figure same as given above. Three to five applications are required where there is water pressure to be overcome.

Liquid Membrane Waterproofing

Liquid membrane (elastomeric) waterproofing is a cold, fluid applied, synthetic rubber, seamless coating system. It is available in single or two-component form and can be applied by trowel or spray. The material forms a continuous, flexible, water impervious film that bonds tightly to a wide range of surfaces, including concrete, stone, masonry, wood, and metal. On a troweled surface, an application rate of five gallons per 125 sq. ft. of surface will produce a cured coating thickness of approximately 55 mils.

Two workmen should coat about 650 sq. ft. per day at the following labor cost:

	Hours	Rate	Total	Rate	Total
Labor	16.0	$....	$....	$16.46	$263.36
Cost per sq. ft.					.41

07150 DAMPPROOFING

Dampproof Paints for Exterior Concrete or Masonry Surfaces.

Where a heavy head of water is to be overcome, the concrete or masonry should preferably be waterproofed by the membrane method. Where only a dampproofing material is required, there are a number of paints on the market for this purpose. The application of one, two or more coats of paint is intended to make the walls impervious to dampness.

Number of coats required and covering capacity will vary with porosity and surface of wall.

Heavy Dampproof Paints, Per 100 Sq. Ft.

Name of Material	Recommended for use on	No. Cts.	Sq. Ft. per Gal.	Gals. Reqd. 100 Sq. Ft.	Lab. Hrs. Applying
Dehydratine 4	Fndtn. Walls	2	33	3	2.2
Dehydratine 10	Mason Walls	1	30	3⅓	1.2
Hydrocide Semi-Mastic	Fndtn. Walls	1	30–50	3	1.2
	Fndtn. Walls	2	15–18	6	2.2
Hydrocide 600	Fndtn. Walls	1	75–100	1½	1.2
Hydrocide 648	Fndtn. Walls	2	50	2	2.2
Marine Liquid	Below Grade	1	50–75	1¾	1.2
Sikaseal	Fndtn. Walls	2	60–80	1½	2.2
Tremco 110	Fndtn. Walls	1	100	1	1.2

07190 VAPOR BARRIERS/RETARDANTS

The combination of high indoor relative humidities and low outside temperatures causes condensation to form within the structure. Serious damage can result from this moisture in the way of rotting framing members, paint deterioration, wet walls and ceilings, etc.

Vapor barriers are recommended where the above conditions exist.

Vapor seal paper should be installed on the warm side of wall, floor, ceiling, roof, etc. Paper should be installed with joints running parallel to and over framing members. All joints should be lapped about 2 inches. Where paper is to be exposed, nail wood lath strips over paper along framing members to provide a neat and permanent job.

Approximate prices on the various types of vapor seal paper are as follows:

Description	Price per Roll 500 Sq. Ft.
2 sheets 30-lb basis kraft paper cemented together with asphalt	$ 5.00
2 sheets asphalt coated 30-lb kraft paper cemented together with asphalt and reinforced with jute cords	$ 9.00
1 sheet heavy kraft paper coated two sides with a reflective surface	$12.00

Sisalkraft.—A strong, waterproof, windproof building paper consisting of 2 sheets of pre-treated kraft paper cemented together with 2 layers of special asphalt and reinforced with 2 layers of crossed sisal fibers. Costs 2 cts. per sq. ft.

Copper-Armored Sisalkraft is a combination of Sisalkraft and Anaconda electro-sheet copper bonded under heat and pressure. It is available in 3 weights, 1, 2 or 3 oz. copper per sq. ft.

It may be used for waterproofing, flashings, ridge roll flashing, etc. Approximate prices are as follows: 1-oz. costs 30 cts. per sq. ft.; 2-oz. costs 50 cts. per sq. ft. and 3-oz. costs 70 cts. per sq. ft. Above prices subject to discount on large orders.

Polyethylene film comes in 100' rolls of various widths in thicknesses of .002", .004", .006", and .008". Costs vary from $1\frac{1}{2}$¢ to 4¢ per sq. ft. Tape for sealing joints costs $3.25 for a roll 2" wide by 100' long in the .004" thickness.

Labor Placing Vapor Seal Paper.—Where vapor seal paper is applied to the interior of exterior stud walls, using a stapling machine, to prevent the penetration of moisture and condensation, a carpenter should handle and place 2,000 to 2,500 sq. ft. per 8-hr. day, at the following labor cost per 100 sq. ft.:

	Hours	Rate	Total	Rate	Total
Carpenter	0.4	$....	$....	$16.47	$6.59
Cost per sq. ft.		....			.07

07200 INSULATION

Most batt and blanket materials have "R" values of around 3.5 to 3.7 per inch of thickness. 3" units will provide an "R" of 11, $3\frac{1}{2}$" of 13 and 6" of 19. The 6" units are commonly used in attics and fitted between the joists; but more and more they are used in walls with 2x6 studs placed 24" o.c. in place of the standard 2x4s 16" o.c. If foil faced units are used and an air space of at least ³⁄₄" maintained between the foil and the warm side of the room the "R" rating can be increased some 15%.

Because these materials are light and precut to fit standard construction one carpenter can handle the installation. Friction fit type, with no attachments, will install at the rate of 2000 sq. ft. per day. Flanged type bound in kraft paper or foil will install at 1800 sq. ft. per day.

Labor Cost of 100 sq. ft. Faced Insulation

	Hours	Rate	Total	Rate	Total
Carpenter	.45	$....	$....	$16.47	7.41
Cost per sq. ft.					.075

07211 LOOSE FILL INSULATION

Number of Sq. Ft. of Surface Covered By One Bag of Loose Insulating Wool Weighing 40 Lbs. and Containing 4 Cu. Ft. Not Including Area Covered By Studs or Joists

Density per Cu. Ft. Lbs.	Actual Cu. Ft.	Thickness of Loose Insulating Wool in Inches					
		1	2	3	3½	3⅝	4
6	6.67	80.0	40.0	26.6	22.8	22.0	20.0
7	5.72	68.5	34.4	22.9	19.7	18.8	17.2
8	5.00	60.0	30.0	20.0	17.2	16.6	15.0
9	4.45	53.4	26.7	17.8	15.2	14.7	13.4
10	4.00	48.0	24.0	16.0	13.7	13.3	12.0

Labor Placing Loose or Bulk Insulating Wool.—When placing loose or bulk insulating wool between wood studs, a man should place about 350 to 450 sq. ft. per 8-hr. day, at the following labor cost per 100 sq. ft. :

	Hours	Rate	Total	Rate	Total
Labor	2	$....	$....	$16.47	$32.94
Cost per sq. ft.			....		.33

Vermiculite Or Perlite Loose-fill Building Insulation.—Vermiculite loose- fill insulation is used to insulate attics, lofts, and side walls. It is fireproof, not merely "fire resistant." The fusion point is 2200° to 2400° F.

It is completely mineral and does not decompose, decay, or rot. Because of its granular structure, vermiculite is free-flowing, assuring a complete insulation job without joints or seams. Rodents cannot tunnel into it, and it does not attract termites or other vermin. Vermiculite is a non-conductor and is excellent protection around electrical wiring.

Vermiculite loose-fill is marketed in 4 cu. ft. bags. Approximate coverage per bag, based on joists spaced 16" o.c., is 14 sq. ft. for a 3⅝" thickness; 9 sq. ft. for 5½" thickness. A bag will cost about $3.60.

Approximate Coverage of Vermiculite
Fill in Cavity and Block Walls

Sq. ft. wall area	Quantity of Bags Required				
	1" cavity	2" cavity	2½" cavity	8" block	12" block
100	2	4	5	7	13
500	10	20	25	34	63
1,000	21	42	50	69	125
5,000	104	208	250	545	625
10,000	208	416	500	1,090	1,250

*1 bag=4 cu. ft.

07212 RIGID INSULATION

Rigid insulation boards are made of expanded polystyrene, often called "beadboard"; extruded polystyrene sold as Styrofoam; urethane; glass fibers; glass foams; and wood and vegetable fibers. Many rigid boards are made to serve dual purposes such as sheathing, lath and even interior finish. Many are also used for roof decks and perimeter insulation which are discussed separately below.

Urethane boards have the highest "R" value per inch of thickness, 7.14, but are flammable and must be covered over, or treated. The cost is around 25¢ per sq. ft. plain, 33¢ treated. The sizes available are many including the standard 4' x 8' sheet and in thicknesses from ½" to 24". Because of its high insulating value per thickness it is ideal for use in cavity walls and for applying to the inside face of an exterior masonry wall. In cavity walls it is applied against the outside face of the inside wythe leaving a clear air space for any penetrating water to drain out.

Where urethane board is used against exterior masonry walls the surface must be clean and even. Nailers must be applied wherever wood trim will be added, such as at the base and around doors and windows. The board is then applied to the wall with sufficient mastic to cover at least 50% of the back when shoved in place. Some manufacturers supply special channel systems for mechanical attachment as an alternative. The urethane board can then be covered with plasterboard or may be plastered direct. Because of the fire hazard, which ever method is chosen, the interior finish coat must be run up above the finished ceiling to cover all portions of the urethane as it is the fireproofing for it.

Extruded polystyrene, or Styrofoam under which name it is marketed by Dow Chemical Company, also has a high "R" value, 5.4 per inch of thickness. It costs about 29¢ per sq. ft. in 1" thickness. It is installed in the same locations and by the same methods as the urethane board above, and like urethane board, it should have a continous interior finish that will protect it from fire.

Polystyrene which is molded is known as beadboard. It has a lower "R" rating, 3.85, and is absorbent, but it is much cheaper, about 12¢ per sq. ft. in 1" thickness. It too must be covered with a fireproofing finish.

Glass fiber boards come in various densities. 1" board in a medium weight gives an "R" of 4.35 and costs around 30¢ per sq. ft.

The stronger rigid boards may also be used for sheathing on wood frame construction. But a 1" board in combination with standard 3½" batts will give a wall with the highly desirable rating of R19 or better and may be preferrable to using 6" batts and 6" studs to gain such a rating.

Rigid board is light, easily cut and can be installed by one man who should erect about 1,500 sq. ft. per day depending on the amount of fitting.

Labor Cost of 100 sq. ft. of 1" Rigid Board
Applied to a Masonry Wall

	Hours	Rate	Total	Rate	Total
Mechanic	.54	$....	$....	$16.47	$8.90
Cost per sq. ft.					.09

07240 ROOF AND DECK INSULATION

Roof and deck insulation is often of the same material as has been discussed under rigid board insulation but the units are generally furnished in smaller sizes more suitable for handling in exposed conditions. In addition to glass fiber, perlite, urethane and polystyrene boards, wood and mineral fiber boards costing as little as 25¢ per sq. ft. per 1" of thickness, and foam glass blocks are often specified. Foamglas is a unique material which is inorganic, incombustible, dimensionally stable, water and vapor proof and weighs only 8.5# per cubic foot but is strong enough to be used under parking and promonade decks. It can also be ordered with an 1/8 " per foot slope. Minimum thickness is 1½" and is available in ½" increments up to 4". 1½" material has an "R" value of 4.2 and costs about 60¢ per sq. ft. in 1½" thickness. One roofer will install some 800 sq. ft. per day at the following labor cost:

Labor Cost of 100 sq. ft. of 1" Deck Insulation

	Hours	Rate	Total	Rate	Total
Roofer	1.0	$....	$....	$17.18	$17.18
Cost per sq. ft.					.17

07250 PERIMETER INSULATION

Board forms of polystyrene, urethane, fiberglass and cellular glass are commonly employed as insulation on foundation walls and under slab edges to insulate an on- grade floor construction at the outer wall. Material costs are from 15¢ to 25¢ per sq. ft. in 1" thickness.

As perimeter insulation must fit the outline of the interior face of the outside wall and must be run continuously around offsets, perimeter ducts, column foundations and other irregularities, unit costs will vary widely. In general, insulation is run down 24" below outside grade. In addition, it may be run back 24" under the outer edge of the slab. One carpenter should apply around 800 sq. ft. per day on straight run work, at the following labor cost per 100 sq. ft.:

	Hours	Rate	Total	Rate	Total
Carpenter	1.14	$....	$....	$16.47	$16.47
Cost per sq. ft.					.17

07300 ROOF SHINGLES AND ROOFING TILES

Roofing is estimated by the square, containing 100 sq. ft. The method used in computing the quantities will vary with the kind of roofing and the shape of the roof.

The labor cost of applying any type of roofing will be governed by the pitch or slope of the roof, size, plan of same (whether cut up with openings, such as skylights, penthouses, gables, dormers, etc.), and upon the distance of the roof above the ground, etc.

A Short Method of Figuring Roof Areas

To obtain the number of square feet of roof area, where the pitch (rise and run) of the roof is known, take the entire flat or horizontal area of the roof and multiply by the factor given below for the roof slope applicable and the result will be the area of the roof.

Always bear in mind, the width of any overhanging cornice must be added to the building area to obtain the total area to be covered.

Pitch of Roof	Rise and Run	Multiply Flat Area by	Lin. Ft. of Hips or Valleys per Lin. Ft. of Common Run
1/12	2 in 12	1.014	1.424
1/8	3 in 12	1.031	1.436
1/6	4 in 12	1.054	1.453
5/24	5 in 12	1.083	1.474
1/4	6 in 12	1.118	1.500
7/24	7 in 12	1.158	1.530
1/3	8 in 12	1.202	1.564
3/8	9 in 12	1.250	1.600
5/12	10 in 12	1.302	1.612
11/24	11 in 12	1.357	1.685
1/2	12 in 12	1.413	1.732

Hips and Valleys.—The length of hips and valleys, formed by intersecting roof surfaces, running perpendicular to each other and having the same slope is also a function of the roof rise and run. For full hips or valleys, i.e. where both roofs intersect for their full width, the length may be determined by taking the square root of the sum of the rise squared plus twice the run squared.

Using the factors given in the last column of the above table, the length of full hips or valleys may be obtained by multiplying the total roof run from eave to ridge, (not the hip or valley run), by the factor listed for the roof slope involved.

07311 ASPHALT SHINGLES

Estimating Quantities of Asphalt Shingles.—Asphalt shingles are sold by the square containing sufficient shingles to cover 100 sq. ft. of roof.

When measuring roofs of any shape, always allow one extra course of shingles for the "starters" at the eaves, as the first or "starting" course of shingles must always be doubled.

Obtain the number of lin. ft. of hips, valleys and ridges to be covered with asphalt shingles and compute same as 1'-0" wide.

Asphalt shingles must be properly nailed—6 nails to a strip, and nailed low enough on the shingle (right at the cut-out), otherwise this will blow off the roof.

Most manufacturers produce a shingle, designed for high wind areas, that interlock in such a manner that all of the shingles are integrated into a single unit. Interlocking shingles are available in single coverage for re-roofing and double coverage for new construction.

Self-sealing shingles with adhesive tabs are produced by most manufacturers.

Nails Required for Asphalt Shingles.—When laying individual asphalt shingles, use a 12 ga. aluminum nail, 1½" long, with a 7/16" head. For laying over old roofs, use nails 1¾" long.

When laying square butt strip shingles, use 11 ga. aluminum nails, 1" long, with a 7/16" head. For laying over old roofs, use nails 1¼" long.

It will require about 1 lb. of nails per sq. of shingles laid.

Sizes and Estimating Data on Asphalt Shingles

Kind of Shingle	Size	No. Shingles per Sq.	Expos. Inches	Length Nails	No. Nails per Shingle	Lbs. Nails per Sq.
3-in.-1 strip	12"x36"	80	5	1	4	1

Approximate Prices on Mineral Surfaced Asphalt Shingles, Seal Tab

Kind of Shingle	Size	Weight per Square	Price per Square
3-in-1 strip	12"x36"	235 lbs.	$26.00
3-in-1 strip	12"x36"	340 lbs.	55.00
3-in-1 strip	12"x36"	350 lbs.	65.00

A roofer will lay one square (100 sq. ft.) of asphalt shingles in 1 hour on double pitched roofs with no hips, valleys or dormers. Figure about 35 lin. feet per hour for fitting hips, valleys and ridges. Hip and ridge roll material will run around 16 cts. per lin. ft.; valley, 30¢ per lin. ft.

Labor Cost of 100 Sq. Ft. (1 Sq.) Strip Asphalt Shingles on Plain Double Pitch or Gable Roofs

	Hours	Rate	Total	Rate	Total
Carpenter	1.0	$....	$....	$16.47	$16.47

On roofs having gables, dormers, etc. add 0.2-hr. time per sq. On difficult constructed hip or English type roofs, add 0.5-hr. time per sq.

07313 WOOD SHINGLES AND SHAKES

The labor cost of laying wood shingles will vary with the type of roof, whether a plain gable roof, a steep roof, or one cut up with gables, dormers, etc. Also with the manner in which they are laid, whether with regular butts, irregular or staggered butts, or thatched butts.

Estimating the Quantity of Wood Shingles.—Ordinary wood shingles are furnished in random widths, but 1,000 shingles are equivalent to 1,000 shingles 4" wide.

Dimension shingles are sawed to a uniform width, being either 4", 5" or 6" wide.

Wood shingles are usually sold by the square based on sufficient shingles to lay 100 sq. ft. of surface, when laid 5" to the weather, 4-bundles to the square.

When estimating the quantity of ordinary wood shingles required to cover any roof, bear in mind that the distance the shingles are laid to the weather makes considerable difference in the actual quantity required.

Number of Shingles and Quantity of Nails Required Per 100 Sq. Ft. of Surface

Distance Laid to Weather	Area Covered by One Shingle Sq. In.	Add for Waste Per Cent	Actual No. per Square Without Waste	Number per Square With Waste	No. of 4-Square Bundles Required	Pounds 3d Nails Req'd
4 "	16	10	900	990	5.0	3.2
4½"	17	10	850	935	4.7	2.8
4½"	18	10	800	880	4.4	2.5
5 "	20	10	720	792	4.0	2.0
5½"	22	10	655	720	3.6	1.6
6 "	24	10	600	660	3.3	1.5

Labor Laying One Square (100 Sq. Ft.) Wood Shingles

Class of Work	Mechanic	Number Laid per 8-Hr. Day	Distance Shingles are Laid to Weather					
			4"	4 1/4"	4 1/2"	5"	5 1/2"	6"
Plain Gable or Hip Roofs	Carpenter	2,000–2,200	3.8 hrs.	3.6 hrs.	3.4 hrs.	3.0 hrs.	2.8 hrs.	2.5 hrs.
	Shingler	2,750–3,000	2.8	2.6	2.5	2.3	2.0	1.9
Difficult Gable Roofs, cut up with gables, dormers, hips, valleys, etc.	Carpenter	1,700–1,900	4.5	4.2	4.0	3.6	3.3	3.0
	Shingler	2,200–2,500	3.4	3.2	3.0	2.7	2.5	2.3
Difficult Hip Roofs, Steep English Roofs, hips, valleys, etc.	Carpenter	1,300–1,500	5.7	5.4	5.1	4.6	4.2	3.8
	Shingler	2,000–2,200	3.8	3.6	3.4	3.0	2.8	2.5
Shingles Laid Irregularly or with Staggered Butts on Plain Roofs.	Carpenter	1,700–1,900	4.5	4.2	4.0	3.6	3.3	3.0
	Shingler	2,400–2,700	3.1	3.0	2.8	2.5	2.3	2.1
Shingles Laid Irregularly or with Staggered Butts on Difficult Constructed Roofs.	Carpenter	1,100–1,300	6.7	6.3	6.0	5.4	4.9	4.5
	Shingler	1,600–1,800	4.7	4.5	4.2	3.8	3.4	3.1
Shingles with Thatched Butts.*	Shingler	800–1,000	8.9	8.4	8.0	7.1	6.5	6.0
Plain Sidewalls.	Carpenter	1,300–1,500	5.7	5.4	5.1	4.6	4.3	3.8
	Shingler	1,700–1,900	4.3	4.3	4.0	3.6	4.2	3.0
Difficult Sidewalls, having bays, windows, breaks, etc.	Carpenter	1,100–1,250	6.8	6.4	6.0	5.4	5.0	4.5
	Shingler	1,400–1,650	5.2	5.0	4.6	4.2	3.8	3.5

To obtain number of bundles of shingles required, divide number of shingles as given above by 200, and the result will be the number of bundles required, i.e, 2,200 shingles ÷ 200 equals 11 bundles; 1,000 ÷ 200=5 bundles; 800 ÷ 200=4 bundles, etc.

*Shingles with thatched butts require 25% more shingles than when laid regularly.

+Use "4" column for carpenter or shingler time per 1,000 shingles. (5 bundles.)

The above table is based on using 2 nails to each shingle and 10 per cent waste.

07314 SLATE ROOFING

Roofing slate is furnished in a number of sizes, thicknesses and finishes to meet the architectural requirements of the buildings on which they are to be used. The four principal classifications are : standard slate ; textural slate ; graduated slate and flat slate roofs.

Standard Roofs.—Standard slate roofs are composed of slate approximately 3/16" thick (commercial standard slate), of one uniform standard length and width; also one length and random widths or random lengths and widths. Any width less than one-half the length is not recommended. Standard roofs are suitable for any building where a permanent roofing material is desired at a minimum cost. If desired, the butts or corners may be trimmed to give a hexagonal, diamond or "Gothic" pattern for all or part of the roof.

Textural Roofs.—The term "textural" is used to designate slate of rougher texture than the standard, with uneven tails or butts, and with variations of thickness or size. In general, this term is not applied to slate over ⅜" in thickness.

Graduated Roofs.—The graduated roof combines the artistic features of the textural slate roof with additional variations in thickness, size, and exposure. The slate is so arranged on the roof that the thickest and longest occur at the eaves and gradually diminish in size and thickness until the ridges are reached. Slate for roofs of this type can be obtained in any combination of thicknesses from 3/16" to 3/4" and heavier when especially desired.

In addition to the usual standard sizes, slate above ½" thick are produced in lengths up to 24". The graduations in lengths generally range from 24" to 12". The variations in length will at once provide a graduation in exposure by using the standard 3" lap. To illustrate, a suitable range of lengths might be :

		3/8" thick		14" long		No exposure
Under Eaves Course						
First Course		3/4"	"	24"	"	10½" "
1	"	3/4"	"	24"	"	10½" "
2	Courses	1/2"	"	22"	"	9½" "
2	"	1/2"	"	20"	"	8½" "
2	"	3/8"	"	20"	"	8½" "
4	"	3/8"	"	18"	"	7½" "
5	"	1/4"	"	16"	"	6½" "
3	"	1/4"	"	14"	"	5½" "
3	"	3/16"	"	14"	"	5½" "
8	"	3/16"	"	12"	"	4½" "

Random widths should be used and so laid that the vertical joints in each course are broken and covered by the slate of the course above.

Size of Slate.—The following table gives the different sizes in which roofing slate is furnished, number required per square when laid with 3" lap, and number of nails required per square based on 2 nails to each piece of slate.

Size of Slate Inches	Number per Square	Number of Nails Reqd. per Square	Size of Slate Inches	Number per Square	Number of Nails Reqd. per Square
10x 6	686	1,372	16x12	185	370
10x 7	588	1,176	16x14	159	318
10x 8	515	1,030	18x 9	213	426
10x10	414	828	18x10	192	384
12x 6	533	1,066	18x11	175	350
12x 7	457	914	18x12	160	320
12x 8	400	800	18x14	137	274
12x 9	355	710	20x10	169	338
12x10	320	640	20x11	154	308
12x12	267	534	20x12	141	282
14x 7	374	748	20x14	121	242
14x 8	327	654	20x16	106	212
14x 9	290	580	22x11	138	276
14x10	261	522	22x12	126	252
14x12	218	436	22x14	109	218
14x14	188	376	22x16	95	190
16x 8	277	554	24x12	115	230
16x 9	246	492	24x14	98	196
16x10	221	442	24x16	86	172

Slate ³⁄₄" and thicker and 24" or over in length should have 4 nail holes.

Weight of Slate Roofing.—A square of slate roofing, i.e., sufficient slate to cover 100 sq. ft. of roof surface with a standard 3" lap, will vary from 650 to 8,000 lbs. for thicknesses from the commercial standard 3/16" to 2".

The weight of slate varies with the size of the slate, color and quarry, and even sometimes in the same quarry. The variation may be from 10% above to 15% below the weights given in the following table :

Average Weight of Slate per Square (100 sq. ft.)

	Slate Thickness in Inches	Sloping Roof Allowing for 3" Lap	Flat Roof Without Lap
Standard	3/16	700	240
Selected full	3/16	750	250
	1/4	900	335
	3/8	1,400	500
	1/2	1,800	675
	3/4	2,700	1,000
	1	4,000	1,330
	11/4	5,000	1,670
	11/2	6,000	2,000
	13/4	7,000	
	2	8,000	

Nails Required for Slate Roofing.—The quantity of nails required for slate roofing will vary with size, kind, etc., but the following table gives quantities of those commonly used :

Approximate Number of Nails to the Pound

Length in Inches	"Copperweld" Slating Nails	Copper Wire Slat. Nails	Cut Copper Slat. Nails	Cut Brass Nails	Cut Yellow-Metal Slat. Nails
1	386	270	...	...	...
1¼	211	144	190	164	154
1½	176	134	135	140	140
1¾	133	112	100	108	...
2	87	104	...	88	...
2¼	...	46	...	80	...
2½	...	...	...	64	...
2¾	...	...	...	52	...
3	...	...	...	48	...

Laying 16"x8" Roofing Slate.—When working on plain roofs that do not require much cutting, 2 slaters and a helper should lay felt, handle and lay 350 to 450 sq. ft. (3½ to 4½ sqs.) of 16"x8" slate per 8-hr. day. On more complicated hip or gable roofs, requiring considerable cutting and fitting for hips, valleys, dormers, etc., 2 slaters and a helper should handle and lay 175 to 225 sq. ft. (1¾ to 2¼ sqs.) of 16"x8" slate per 8-hr. day.

Laying 18"x9" Roofing Slate.—When laying 18"x9" slate on plain roofs, 2 slaters and a helper should lay felt, handle and lay 400 to 500 sq. ft. (4 to 5 sqs.) per 8-hr. day.

On more complicated hip or gable roofs, requiring considerable cutting and fitting for hips, valleys, dormers, etc., 2 slaters and a helper should lay 250 to 300 sq. ft. (2½ to 3 sqs.) per 8-hr. day.

Laying 20"x10" Roofing Slate.—When laying 20"x10" slate on plain roofs, 2 slaters and a helper should lay felt, handle and lay 500 to 600 sq. ft. (5 to 6 sqs.) of roof per 8-hr. day, but on more complicated roofs, requiring considerable cutting and fitting for hips, valleys, dormers, etc., the same crew would lay only 300 to 350 sq. ft. (3 to 3½ sqs.) per 8-hr. day.

Laying 22"x12" Roofing Slate.—On straight roofs, 2 slaters and a helper should lay felt, handle and lay 550 to 650 sq. ft. (5½ to 6½ sqs.) of 22"x12" slate per 8-hr. day, but on the more complicated roofs, requiring cutting and fitting for hips, valleys, dormers, etc., the same crew would lay only 350 to 400 sq. ft. (3½ to 4 sqs.) per 8-hr. day.

Laying Graduated Roofing Slate.—When laying graduated roofing slate from 3/16" to 3/4" thick and 12" to 24" long, considerable care is required in selecting the right slate for each course and laying them to obtain the desired effect. On work of this kind 2 slaters and a helper should lay felt, handle and lay 150 to 200 sq. ft. (1½ to 2 sqs.) of roof per 8-hr. day.

07400 PREFORMED ROOFING & SIDING

Corrugated steel is used extensively for roofing and siding steel mills, manufacturing plants, sheds, grain elevators and other industrial structures.

It is made with various corrugations, varying in width and depth, but the 2½" corrugation width is the most commonly used.

The sheets are usually furnished 26" wide and 6'-0" to 30'-0" long. Some manufacturers offer siding protected with vinyl coatings.

When estimating quantities of corrugated siding or roofing, always select lengths that work to best advantage and be sure and allow for both end and side laps.

Corrugated siding or roofing is nailed to the wood framework or siding when used on wood constructed buildings. When used as a wall and roof covering on structural steel framing, the corrugated sheets are fastened to the steel framework, using clip and bolts, which are passed around or under the purlins (which usually consist of channels, angles or Z bars). When angles are used for purlins, clinch nails are sometimes used for fastening the corrugated sheets.

When used for siding one corrugation lap is usually sufficient, but for roofing two corrugations should be used and if the roof has only a slight pitch, the lap should be three corrugations.

When used for siding, a 1" to 2" end lap is sufficient, but when laid on roofs it should have an end lap of 3" to 6" depending upon the pitch of the roof. For a 1/3 pitch a 3" lap is sufficient; for a 1/4 pitch a 4" lap should be used; and for a 1/8 pitch a 5" end lap.

When applying to wood sheathing or strips, the nails should be spaced about 8" apart at the sides. When applied to steel purlins, the side laps should extend over at least 1½ corrugations, and the sheets should be riveted together every 8" on the sides and at every alternate corrugation on the ends.

Number of Sq. Ft. of Corrugated Sheets Required to Cover 100 Sq. Ft. of Surface, Using 26"x96" Sheets

2½-Inch Corrugation	Length of End Lap in Inches 1	2	3	4	5	6
Width	Square Feet of Corrugated Metal					
Side lap, 1 corrugation	110	111	112	113	114	115
Side lap, 1½ corrugations...	116	117	118	119	120	121
Side lap, 2 corrugations.......	123	124	125	126	127	128
Side lap, 2½ corrugations...	130	131	132	133	134	135
Side lap, 3 corrugations.......	138	139	140	141	142	143

Placing Corrugated Steel Roofing or Siding on Wood Framing.—If cor- rugated sheets used for roofing or siding are nailed to wood strips or framing, a man and helper should place

100 sq. ft. (1 sq.) of 26"x96" or larger, in ⅞ to 1⅛ hr. or 7 to 9 sqs. per 8-hr. day, at the following labor cost per 100 sq. ft. (1 sq.) :

Labor Cost of 100 Sq. Ft. (1 Sq.) Corrugated Metal Roofing on Wood Framing

	Hours	Rate	Total	Rate	Total
Carpenter	1	$....	$....	$16.47	$16.47
Labor	1	....	....	12.54	12.54
Cost per 100 sq. ft.			$....		$29.01

Labor Cost of 100 Sq. Ft. (1 Sq.) Corrugated Metal Siding on Wood Framing

	Hours	Rate	Total	Rate	Total
Carpenter	1.2	$....	$....	$16.47	$19.76
Labor	1.2	....	....	12.54	15.05
Cost per 100 sq. ft.			$....		$34.81

Placing Corrugated Steel Roofing on Steel Framing.—The labor cost of applying corrugated steel roofing to structural steel framing where the sheets are fastened with clips or bolts, will vary with the size of the roof area, regularity of roof, height above ground, etc., but on a straight job, 2 sheeters and 2 laborers should place 550 to 650 sq. ft. per 8-hr. day, at the following labor cost per 100 sq. ft. :

	Hours	Rate	Total	Rate	Total
Labor handling, hoisting	2.67	$....	$....	$12.54	$33.48
Sheeters	2.67	....	....	17.70	47.26
Cost per 100 sq. ft.			$....		$80.74

On irregular roofs, above costs may be increased 25 to 50%.

Placing Corrugated Steel Siding on Steel Framing.—The labor cost of applying corrugated steel siding to walls will vary greatly according to length and height of walls, quantity and regularity of openings, etc., but on a straight job, 2 sheeters and 2 laborers should place 450 to 550 sq. ft. per 8-hr. day, at the following labor cost per 100 sq. ft. :

	Hours	Rate	Total	Rate	Total
Labor handling, hoisting	3.2	$....	$....	$12.54	$40.13
Sheeters	3.2	....	....	12.70	56.64
Cost per 100 sq. ft.			$....		$96.77

On coal tipples, grain elevators and other structures having high and irregular walls, the above costs may be doubled.

07415 CORRUGATED ALUMINUM ROOFING AND SIDING

When using aluminum roofing and siding, it is very important when applying the sheets that they be insulated against electrogalvanic action. Ordinarily aluminum has a high resistance to corrosion but when it is contiguous to steel or copper in the presence of moisture, an electrolytic cell is formed and the resulting flow of current dissolves the aluminum, causing severe pitting. This condition can be avoided by preventing metal-to- metal contact.

To provide adequate drainage, the roof surface should never have a slope less than 2½ in. per ft. and preferably not less than 3 in. per ft.

For roofing, sheets should have a side lap of 1½ corrugations. For siding, sheets should be lapped 1 corrugation.

A 6 in. overlap at the ends is recommended for roofing and 4 in. for siding.

Sizes, Weights and Coverage Data on Corrugated Aluminum Sheets Used for Roofing

Lengths 5'-0" to 12'-0" in 6" increments. Widths, 35" and 48⅜". Thicknesses, .024" and .032". Corrugation, 2.67" pitch and ⅞" deep. Coverage, 32" for 35" width and 45⅜" for 48⅜".

Data on .032" Corrugated Aluminum Roofing Sheets

Sheet	Square Feet		Weight per Sheet in		Approx. No. Sheets* per 100 Sq. Ft.	
Length in Feet	per Sheet		Lbs.			
	35"	48⅜"	35"	48⅜"	35"	48⅜"
5	14.58	20.16	8.04	11.03	6.86	4.96
5½	16.04	22.17	8.85	12.13	6.23	4.51
6	17.50	24.19	9.65	13.24	5.71	4.13
6½	18.96	26.20	10.46	14.34	5.27	3.82
7	20.42	28.22	11.26	15.44	4.90	3.54
7½	21.88	30.23	12.07	16.55	4.57	3.31
8	23.33	32.25	12.87	17.65	4.29	3.10
8½	24.79	34.27	13.68	18.75	4.03	2.92
9	26.25	36.28	14.48	19.86	3.81	2.76
9½	27.71	38.30	15.29	20.96	3.61	2.61
10	29.17	40.31	16.09	22.06	3.43	2.48
10½	30.63	42.33	16.90	23.16	3.26	2.36
11	32.08	44.34	17.70	24.27	3.12	2.26
11½	33.54	46.36	18.51	25.37	2.98	2.16
12	35.00	48.37	19.31	26.47	2.86	2.07

For .024" thickness use data given above except reduce weights 25 percent.

*For side and end lap allowance, add approximately 16% for 35" width and 12% for 48⅜" width.

Sizes, Weight and Coverage Data on Corrugated Aluminum Sheets Used for Siding

Lengths, 5'-0" to 12'-0" in 6" increments. Widths, 33¾" and 47⅛". Thicknesses, .024" and .032". Corrugation, 2.67" pitch and ⅞" deep. Coverage, 32" for 33¾" width and 45⅜ for 47⅛" width.

Data on .032" Corrugated Aluminum Siding Sheets

Sheet Length in Feet	Square Feet per Sheet 33¾"	Square Feet per Sheet 47⅛"	Weight per Sheet in Lbs. 33¾"	Weight per Sheet in Lbs. 47⅛"	Approx. No. Sheets* per 100 Sq. Ft. 33¾"	Approx. No. Sheets* per 100 Sq. Ft. 47⅛"
5	14.06	19.64	7.76	10.75	7.11	5.09
5½	15.47	21.60	8.54	11.82	6.46	4.63
6	16.87	23.56	9.31	12.90	5.93	4.24
6½	18.28	25.53	10.09	13.97	5.47	3.92
7	19.69	27.49	10.87	15.05	5.08	3.64
7½	21.09	29.45	11.65	16.12	4.74	3.40
8	22.50	31.42	12.42	17.20	4.44	3.18
8½	23.91	33.38	13.20	18.27	4.18	3.00
9	25.31	35.34	13.97	19.35	3.95	2.83
9½	26.72	37.31	14.75	20.42	3.74	2.68
10	28.12	39.27	15.52	21.50	3.56	2.55
10½	29.53	41.23	16.30	22.57	3.39	2.43
11	30.94	43.20	17.08	23.65	3.23	2.31
11½	32.34	45.16	17.86	24.72	3.09	2.21
12	33.75	47.13	18.63	25.80	2.96	2.12

For .024" thickness use data given above, except reduce weights 25 percent.

*For side and end lap allowance, add approximately 9% for 33¾" width and 6% for 47⅛" width.

Data on Nails Required for Roofing

Kind of Nail	Length Inches	Number per Lb.	Price per Lb.
Galvanized Needle Point	1¾"	83	$ .75
Galvanized Needle Point	2 "	77	.75
Aluminum Nails—Neoprene Washers	1¾"	318	2.80
Aluminum Nails—Neoprene Washers	2 "	285	2.80

Labor Placing Corrugated Aluminum Roofing and Siding.

Labor Cost of 100 Sq. Ft. (1 Sq.) Corrugated Aluminum
Roofing on Wood Framing.
  Based on using sheets 35" wide, .032" thick, weighing .552
lbs. per sq. ft.

	Hours	Rate	Total	Rate	Total
Carpenter	1	$....	$....	$16.47	$16.47
Labor	1	....	....	12.54	12.54
Cost per 100 sq. ft.			$....		$29.01

Cost of 100 Sq. Ft. (1 Sq.) Corrugated Aluminum Siding
on Wood Framing
  Based on using sheets 33¾" wide, .032" thick, weighing .552
lbs. per sq. ft.

	Hours	Rate	Total	Rate	Total
Carpenter	1.2	$....	$....	16.47	$19.76
Labor	1.2	....	....	12.54	15.05
Cost per 100 sq. ft.			$....		$34.81

Cost of 100 Sq. Ft. (1 Sq.) Corrugated Aluminum Siding
on Steel Framing, Straight Work
  Based on using sheets 33¾" wide, .032" thick, weighing .552
lbs. per sq. ft.

	Hours	Rate	Total	Rate	Total
Labor handling-hoisting	3.2	$....	$....	12.54	$40.13
Sheeters	3.2	....	....	17.70	56.64
Cost per 100 sq. ft.			$....		$96.77

Add extra labor for high and irregular walls such as grain
elevators, conveyor housing, etc. Add for closures, flashings,
etc.

Cost of 100 Sq. Ft. (1 Sq.) Corrugated Aluminum Roofing
on Steel Framing, Straight Work
  Based on using sheets 35" wide, .032" thick, weighing .552
lbs. per sq. ft.

	Hours	Rate	Total	Rate	Total
Labor handling-hoisting	2.67	$....	$....	12.54	$33.48
Sheeter	2.67	....	....	17.70	47.26
Cost per 100 sq. ft.			$....		$80.74

Add extra labor for high or irregular roof areas. Add for closures, flashings, etc.

07460   CLADING/SIDING

Wood Siding is available in plain bevel, plain shiplap, and tongue and groove patterns to be used horizontally in heights from 4" to 12". Bevel siding is either ½" or ¾" thick, T&G, and shiplap nominal 1". Cedar is the most common wood for siding but redwood, fir, hemlock and spruce are also stocked. Wood may also be installed vertically with wood battens at the joints. Bevel siding ½" thick in clear "A" grade cedar will run around $1.10 a board foot; in redwood around $1.25. T&G red cedar 1x8s "D" and better will run 80¢ a board foot; T&G redwood, clear and better, runs $1.40.

Hardboard Siding, 7/16" thick and factory primed, will run 35¢ a square foot.

Plywood Siding in 4x8 sheets, rough textured and grooved to look like individual boards, ⅝" thick, will run around 75¢ a square foot in select grade Douglas fir, $1.10 in redwood.

Aluminum Siding factory finished white and in 8" widths will run 50¢ uninsulated, 85¢ insulated per sq. ft. Flashing strips will cost 20¢ a foot; inside corners 45¢; outside corners 90¢ a lineal foot. Colored nails cost $2.25 a pound.

Vinyl Siding, factory finished in 8" wide strips will cost about 60¢ a square foot, uninsulated, 80¢ with insulating backup. Corner boards will run 80¢ for outside positions, 45¢ for inside. Trim moldings will run 23¢ per lin. ft., while soffits will run 90¢ a sq. ft.

Labor Placing Drop Siding

Measured Size Inches	Actual Size Inches	Class of Workmanship	Feet B.M. Placed per 8-Hr. Day	Carpenter Hours per 1000 Ft. B.M.
6	5¼	Rough Ends	525–575	14.5*
6	5¼	Fitted Ends	350–400	21.4
6	5¼	Mitered Corners	285–325	26.3
8	7¼	Rough Ends	600–650	12.8*
8	7¼	Fitted Ends	415–460	18.2
8	7¼	Mitered Corners	325–375	23.0

*Where an electric saw is used to square both ends of the siding before placing, leaving the exposed corners rough to be covered with metal corner pieces, deduct 1 to 1½ hours time per 1,000 ft. b.m. from the time given above.

Quantity of Bevel Siding Required Per 100 Sq. Ft. of Wall

Measured Size Inches	Actual Size Inches	Exposed to Weather	Pattern	Add for Lap	Ft. B.M. Req. per 100 Sq. Ft. Surface
½ x 4	½ x 3¼	2¾	Regular	46%	151
½ x 5	½ x 4¼	3¾	Regular	33%	138
½ x 6	½ x 5¼	4¾	Regular	26%	131
½ x 8	½ x 7¼	6¾	Regular	18%	123
⅝ x 8	⅝ x 7¼	6¾	Regular	18%	123
¾ x 8	¾ x 7¼	6¾	Rabbetted	18%	123
⅝ x10	⅝ x 9¼	8¾	Rabbetted	14%	119
¾ x10	¾ x 9¼	8¾	Rabbetted	14%	119
¾ x12	¾ x11¼	10¾	Rabbetted	12%	117

The above quantities include 5% for end cutting and waste.

Quantity of Drop Siding Required Per 100 Sq. Ft. of Wall

Measured Size Inches	Actual Size Inches	Exposed to Weather	Add for Lap	Ft. B.M. Req. per 100 Sq. Ft. Surface
1x6	¾ x5¼	5¼	14%	119
1x8	¾ x7¼	7¼	10%	115

The above quantities include 5% for end cutting and waste.

Quantity of Shiplap Required Per 100 Sq. Ft. of Surface

Measured Size Inches	Actual Size Inches	Add for Lap	Ft. B.M. Req. per 100 Sq. Ft. Surface
1x 8	¾ x7¼	10%	115
1x10	¾ x9¼	8%	113

The above quantities include 5% for end cutting and waste.

Labor Placing Bevel Siding

Measured Size Inches	Actual Size Inches	Exposed to Weather Inches	Class of Workmanship	Feet B.M. Placed per 8-Hr. Day	Carpenter Hours per 1000 Ft. B.M.
4	3¼	2¾	Rough Ends	350–400	21.3**
4	3¼	2¾	Fitted Ends	240–285	30.5
4	3¼	2¾	Mitered Corners	200–240	36.3
5	4¼	3¾	Rough Ends	415–460	18.2**
5	4¼	3¾	Fitted Ends	285–330	25
5	4¼	3¾	Mitered Corners	240–285	30.5
6	5¼	4¾	Rough Ends	475–525	16*
6	5¼	4¾	Fitted Ends	325–375	23
6	5¼	4¾	Mitered Corners	265–310	28
8	7¼	6¾	Rough Ends	570–620	13.4**
8	7¼	6¾	Fitted Ends	375–415	20
8	7¼	6¾	Mitered Corners	300–350	24

Labor Placing Bevel Siding (Con't)

Measured Size Inches	Actual Size Inches	Exposed to Weather Inches	Class of Workmanship	Feet B.M. Placed per 8-Hr. Day	Carpenter Hours per 1000 Ft. B.M.
10	9¼	8¾	Rough Ends	650–700	12**
10	9¼	8¾	Fitted Ends	440–480	17.5
10	9¼	8¾	Mitered Corners	375–425	20
12	11¼	10¾	Fitted Ends	475–525	16
12	11¼	10¾	Mitered Corners	400–450	19

**Where an electric handsaw is used to square both ends of the siding before placing, leaving the exposed corners rough to be covered with metal corner pieces, deduct 1 to 1½ hours time per 1,000 ft. b.m.

07464 PLASTIC SIDING

Fiberglass panels are made to most of the usual configurations and widths and in lengths up to 30', although stock lengths are in the 3' to 12' range. Fiberglass panels may be used to enclose an entire building, walls and roof, or in combination with similarly configured panels of steel, aluminum, protected metal or asbestos. The standard panels are translucent and can be inserted as skylights or windows. They can be either clear or colored, smooth or embossed. Fiberglass has an ignition point of 850-900 degrees. Where greater fire protection is required they may be ordered to be rated "fire retardant" at around 60¢ per square foot extra. The panels can be further treated to resist erosion or corrosion, or reinforced to be resistant to breakage. The weight of the sheet determines the cost. Standard sheets weighing 4 oz. per sq. ft. will cost around 70¢ per sq. ft.; 5 oz., 80¢; 6 oz., 95¢; and 8 oz., $1.25. The material is light and one carpenter should set around 100 sq. ft. per hour at the following cost per 100 sq. ft.;

	Hours	Rate	Total	Rate	Total
Carpenter	1.0	$....	$....	$16.47	$16.47
Cost per sq. ft.			....		.17

07500 MEMBRANE ROOFING

Built-up roofing consists of alternate plies of saturated felt and moppings of pitch with tar saturated felt or asphalt with asphalt saturated felt, covered with a top pouring of pitch or asphalt into which slag or gravel is embedded. On flat roofs with slopes of less than ¼" per foot, on which water may collect and stand, coal tar pitch and felt or a low melting point asphalt bitumen and asphalt felt are generally used. Coal tar pitch is not recommended for roofs having an incline in excess of 1" per foot.

Rules for Measuring Flat Roofs.— When measuring flat roof surfaces that are to be covered with composition, tar and gravel, tin, metal or prepared roofing, the measurements should be taken from the outside of the walls on all four sides to allow for flashing up the side of each wall.

When estimating the area of flat roof surfaces, do not make deductions for openings containing less than 100 sq. ft. and then deductions should be made for just one-half the size of the opening.

Make deductions in full for all openings having an area of 500 sq. ft. or more.

Quantity of Asphalt Required For 100 Sq. Ft of Roof.

Number Plies	Number of Plies Dry	Mopped	Surface to Which Roof is Applied	Lbs. per 100 Sq. Ft.
5	2	3	Wood, plywood, structural wood fiber	120
4	1	3	Poured sypsum, lightweight concrete	120
4		4	Concrete, precast concrete or gypsum	140

For roofing pitch add approximately 5 lbs. per mopping each ply per 100 sq. ft.— add 15 lbs. for top pouring per 100 sq. ft.

The following table gives the quantity of roofing felt required to cover 100 sq. ft. of surface of various thicknesses :

Number of Plies	Add for Waste	No. Sq. Ft. Req'd. 100 Sq. Ft. Roof	Weight per Square of Roof Using No. 15 Felt
1	8%	108	15
2	8%	216	30
3	8%	324	45
4	8%	432	60
5	8%	540	75

The general use of No. 30 felt is as a base sheet over wood decks in 1-ply thickness over which subsequent layers of No. 15 felts are mopped in.

Quantity of Roofing Gravel Required for Built-Up Roofs.—Roofing gravel should be uniformly embedded into a heavy top pouring of asphalt or pitch so that approximately 400 lbs. of gravel or 300 lbs. of slag is used per 100 sq. ft. of roof area.

Labor Applying Built-Up Roofing.—The labor cost of handling materials and applying built-up roofing will vary with the specification used and the type of building to which the roofing is applied. On the low type of building, 1, 2 or 3 stories high, where there is no great distance from the ground to the roof deck, the labor for hoisting materials, and supplying hot asphalt or pitch from the ground to the roof deck is considerably less than on a high building, where there is a greater distance from the kettle to the roof.

A 5-man crew should apply the following number of squares of roof per 8-hr. day, based on first class workmanship throughout:

Type of Roofing	Type "A" Structures	Type "B" Structures
3-ply roof over wood roof deck	24 Sqs.	18 Sqs.
3-ply roof over concrete roof deck or insulation...	22	17
4-ply roof over wood roof deck	22	17
4-ply roof over concrete roof deck or insulation...	18	14
5-ply roof over wood roof deck	18	14

Type "A" buildings are the most convenient type of structure for application of built-up roofing. A low type of building from 1 to 3 stories in height, straight, practically flat area not broken up by very many skylights or variations in the deck elevation.

Type "B" buildings are not convenient type of structures for the application of built-up roofing. These would include high buildings which require considerable handling of materials, and buildings with roofs broken up by skylights or penthouses, sawtooth construction, monitors, or considerable variation in deck levels.

An easy and accurate method of computing the labor cost on any type of roof is to take the cost of a 5-man crew for an 8-hr. day, and divide this by the number of squares of roof applied per day, as follows:

	Hours	Rate	Total	Rate	Total
Foreman (working)............	8	$....	$....	$17.58	$140.64
Roofers (4)............	32	....	....	17.18	549.76
Crew cost per 8-hr. day			$....		$690.40
Cost per sq. 14 sqs. per day........			....		49.31
Cost per sq. 17 sqs. per day........			....		40.61
Cost per sq. 18 sqs. per day........			....		38.36
Cost per sq. 22 sqs. per day........			....		31.38
Cost per sq. 24 sqs. per day........			....		28.77

Material Cost of 100 Sq. Ft. (1 Sq.) 3-Ply Tar and Gravel
Built-Up Roof Over Poured Concrete, Poured Gypsum
Roof Decks, or Insulation Board

Maximum incline 2-in. per ft. Roof consisting of 3 plies of
felt, coal tar pitch surfaced with gravel or slag.

	Rate	Total	Rate	Total
3.3 sqs. No. 15 felt..........	$....	$....	$3.00	$ 9.90
150 lbs. coal tar pitch	....	....	.13	19.50
400 lbs. roofing gravel	....	....	.50	2.00
Fuel, mops, etc................	....	....	3.00	3.00
Cost per 100 sq. ft.		$....		$34.40

Material Cost of 100 Sq. Ft. (1 Sq.) of Each Additional
Ply of No. 15 Felt and Hot Mopping of Pitch

	Rate	Total	Rate	Total
1.1 Sqs. No. 15 Felt........	$....	$....	$3.00	$3.30
25 lbs. Pitch	....	....	.13	3.25
Fuel, Mops, etc.	....	....	.50	.50
Cost per 100 sq. ft.		$....		$7.05

7530-40 ELASTIC SHEET AND FLUID APPLIED ROOFING

These types of elastomeric coatings include neoprene,
Hypalon, urethane, butyl and silicone. Some are applied by
spraying, others in sheet form, and all are subcontracted to
firms licensed by the manufacturer.

Since these coatings are thin, the roof surface to which they
are to be applied must be firm, continuous, smooth, clean, and
dry. New concrete should be sealed with a primer.

Because of their thinness and ability to conform to any shape
and the fact that they can be had in most any colors including
white they are often chosen as the roofing surface for decorative and fluid roof forms. Some materials are also used as traffic
decks.

A 1/16" thick butyl sheet will cost around 65¢ a sq. ft. and
one roofer can install some 225 sq. ft. per day for a total cost of
around $1.25 a sq. ft.

A 1/16" neoprene sheet will have a material cost of some
$1.25 a sq. ft. and will install at the same rate as the butyl for
an installed price of around $1.80 a sq. ft.

Fluid applied Hypalon-neoprene .02" thick will be applied at
the rate of some 100–110 sq. ft. per day and have a material
cost of 85¢ for a total figure of around $2.00 a sq. ft.

07600 FLASHING AND SHEET METAL
07610 SHEET METAL ROOFING

Metal roofing includes galvanized steel, copper, lead, stainless steel and aluminum plus the many combinations and alloys
of these metals such as lead-coated copper, terne (80% lead,
20% tin over copper-bearing carbon steel) microzinc and terne

Labor Placing 100 Sq. Ft. Flat and Standing Seam Tin and Metal Roofing

Description of Roof	Number Sq. Ft. per 8-Hr. Day	Hours per 100 Sq. Ft.		*Combined Labor Cost per 100 Sq. Ft.
		Tinner	Helper	
Flat seam metal roofing, using 24"x96" sheets or larger	300-350	2.5	2.5	$ 80.73
Flat seam tin roofing, using 14"x20" plates	175-225	4.0	4.0	129.16
Flat seam tin roofing, using 20"x28" plates	225-275	3.3	3.3	106.56
Flat seam tin roofing, using 20" tin in rolls	325-375	2.3	2.3	74.27
Flat seam tin roofing, using 14" tin in rolls	275-325	2.8	2.8	90.41
Flat seam tin roofing, using 28" tin in rolls	425-475	1.8	1.8	58.12
Standing seam metal roofing, using 24"x96" sheets	250-300	3.0	3.0	96.87
Standing seam tin roofing, using 14"x20" plates	125-175	5.3	5.3	171.14
Standing seam tin roofing, using 20"x28" plates	175-225	4.0	4.0	129.16
Standing seam tin roofing, using 14" tin in rolls	200-225	3.8	3.8	122.70
Standing seam tin roofing, using 20" tin in rolls	250-300	3.0	3.0	96.87
Standing seam tin roofing, using 24" tin in rolls	275-325	2.8	2.8	90.41
Standing seam tin roofing, using 28" tin in rolls	300-350	2.4	2.4	77.50
V-Crimped metal roofing, using 24"x96" sheets or larger	425-475	1.8	1.8	58.12

*Tinner - $19.75 per hr. and Helper - $12.54 per hr.

coated stainless. Terne and aluminum are the least expensive and copper and lead the costliest. Metal prices tend to fluctuate broadly as many of the ores are imported and at the mercy of the value of the dollar and the political climate of the country they are mined in.

Metal roofs may be applied in many ways. The simplest is the flat seam roof which may be used on slopes as low as ¼" to the foot. Standing seams and battened seams generally need a slope of at least 2½". They are more decorative, and in fact batten designs are often selected solely on their decorative value. Some materials may be ordered with prefabricated battens, others are formed in the traditional way over wood strips. Another decorative roof is the "Bermuda" type, where the metal is applied over wood "steps" provided by the carpenter. The step is based on the width of the metal roll to be used and is sloped at least 2½" to the foot. The roofer interlocks the rolls at each step edge giving a sharp shadowline.

For custom work terne and copper, with or without special coatings, are the usual choices. Many batten roofs today have the batten stamped into the metal and are usually factory finished aluminum or steel.

07620   SHEET METAL FLASHING AND TRIM

Sheet metal work in its many branches is a highly specialized business, because the greater part of the work is performed in the shop and the fabricated materials are sent to the job ready to erect. For that reason an estimate on sheet metal work must take into consideration the cost of the finished materials at the shop or delivered to the building site, plus the labor cost of erection.

When preparing the quantity survey for the different types, shapes, and sizes of metal items, be sure to identify the type and weight, or gauge, of metal that is to be used, e.g., galvanized steel, aluminum, stainless steel, copper, lead, terne, etc. Each one will have its own waste factor, difficulty in shaping, method of jointing, etc.; all affecting the production and installation costs, in addition to the sheet goods cost delivered to the shop.

Placing Hanging Gutter or Eave Trough.—This cost will vary with the slope of the roof, distance above the ground and the method used in hanging the gutter.

On hip or gable roofs having eaves 20'-0" to 25'-0" above the ground, a tinner and helper should place 180 to 220 lin. ft. of gutter per 8-hr. day, at the following labor cost per 100 lin. ft.:

	Hours	Rate	Total	Rate	Total
Sht. met. wkr	4	$....	$....	$19.75	$ 79.00
Labor	4	....	....	12.54	50.16
Cost 100 lin. ft.			$....		$129.16
Cost per lin. ft.			...		1.29

Placing Metal Conductor Pipes or Downspouts.—Metal conductor pipes or downspouts are furnished in both round and square and from 1½" to 6" diameter. The conductor pipe extends down the side of the building from the eave trough to grade, where it is connected to an elbow that throws the water away from the building or is connected with the drainage system. Metal conductor pipe is usually held in place by hooks fastened to the walls.

On one or two story buildings where the conductor pipe is 12'-0" to 25'-0" long, a tinner and helper should place 200 to 250 lin. ft. per 8-hr. day, at the following labor cost per 100 lin. ft.:

	Hours	Rate	Total	Rate	Total
Sht. met. wkr	3.5	$....	$....	$19.75	$ 69.13
Labor	3.5	....	....	12.54	43.89
Cost 100 lin. ft.			$....		$113.02
Cost per lin. ft.			....		1.13

Placing Metal Flashing.—When placing metal roof flashing around parapet walls, etc., it is customary to have the flashing extend under the roofing and up the side of the wall 6" to 12", and then place the counterflashing above to lap over the flashing.

Where it is not necessary to cut reglets in the masonry wall, a tinner should place 140 to 160 lin. ft. of flashing per 8-hr. day, at the following cost per 100 lin. ft.:

	Hours	Rate	Total	Rate	Total
Sht. met. wkr.	5.4	$....	$....	$19.75	$106.65
Cost per lin. ft.			....		1.07

Placing Metal Counterflashing.—When placing metal counterflashing over felt or metal flashing that extends up on the side of the wall, it will be necessary to place the counterflashing in a reglet or open joint in the masonry that can be caulked or sealed to prevent the water getting in back of the counterflashing and under the roof.

Where reglets have previously been cut, as is usually the case on stone copings, requiring only the sealing of the joint, a tinner should place 140 to 160 lin. ft. per 8-hr. day, at the following labor cost per 100 lin. ft.:

	Hours	Rate	Total	Rate	Total
Sht. Met. Wkr.	5.4	$....	$....	$19.75	$106.65
Cost per lin. ft.			....		1.07

Cutting Reglets in Masonry Walls.—The labor cost of cutting reglets in brick walls will depend entirely upon the kind of mortar used and the condition of the mortar at the time of cutting the reglets.

If the brick are laid in portland cement mortar, it will cost more to cut the reglet than when laid in lime mortar. Also, if the reglets are cut within 24 hrs. after the brick are laid, the labor cost will be much less than cutting them two or three weeks later.

If reglets are cut before the mortar has had sufficient time to set, a man should cut 175 to 225 lin. ft. per 8-hr. day, at the following labor cost per 100 lin. ft.:

	Hours	Rate	Total	Rate	Total
Labor	4	$....	$....	$12.54	$50.16
Cost per lin. ft.					.50

If the reglets are cut several days after the brick have been laid, giving the mortar sufficient time to harden, it will require a hammer and chisel to cut out the joints. Under these conditions a man should cut 65 to 80 lin. ft. per 8-hr. day, at the following labor cost per 100 lin. ft.:

	Hours	Rate	Total	Rate	Total
Labor	11	$....	$....	$12.54	$137.94
Cost per lin. ft.			....		1.38

If a portable electric saw, having an abrasive blade or cutting wheel is used for cutting the reglets, a man should cut 10 to 15 lin. ft. of reglet an hour.

Cutting reglets in stone coping or balustrade is usually performed by stone cutters at about the same cost as given above.

Placing Metal Valleys.—When placing metal shingles or valleys in connection with gable or hip roofs, dormers, etc., a tinner should place 115 to 135 lin ft. per 8- hr. day, at the following labor cost per 100 lin. ft.:

	Hours	Rate	Total	Rate	Total
Sht. Met. Wkr.	6.4	$....	$....	$19.75	$126.40
Cost per lin. ft.			....		1.26

Placing Metal Conductor Heads.—On buildings having flat roofs where the conductor pipe or downspout is placed inside of the building, requiring a conductor head at the low point of the roof which is connected to the downspout, a tinner should place one conductor head (made complete in the shop) in 1½ to 2 hrs. at the following labor cost:

	Hours	Rate	Total	Rate	Total
Sht. Met. Wkr.	1.8	$....	$....	$19.75	$35.55

07800 ROOF ACCESSORIES

Plastic Dome Skylights.

This type skylight consists of a thermo-formed acrylic plastic sheet free blown into square, rectangular and circular domes.

This type skylight is furnished as a factory assembled unit; the plastic is mounted in an extruded aluminum frame, and prefabricated curbs are available. The aluminum frame is generally designed to form counterflashing.

Plastic Domes

Size	Cost Per Roof Dome
16"x16"	$ 40.00
24"x24"	50.00
36"x36"	65.00
48"x48"	115.00
24" Round	85.00
48" Round	150.00

Erection cost will vary widely as to the number involved and the preparatory work done by others. As units are vulnerable to breakage and scratching, it takes two men to handle all but the smallest unit. On a job with several units involved two men should set 2 units of up to 10 sq. ft. in a day if the curbing is already in place.

The dome unit is available as an insulating unit which has a secondary acrylic dome mounted under the primary (exterior) dome. This double dome unit provides a dead air space between the domes that reduces heat loss and condensation formation.

ROOF HATCH

A standard size 2'-6"x3'-0" steel roof hatch with insulated curb and cover and hardware will run around $275.00. A mechanic and helper should set this in 2 hrs.

GRAVITY VENTILATORS

12" stationary ventilators of the syphon type with a 24" base will cost around $40 and take two hours to set. Stationary ventilators of the mushroom type with a 24" base will cost $130.00 with the same labor cost.

07900   SEALANTS

With modern construction materials and designs, came the need for good sealants and caulking compounds. No one sealant can solve the requirements across the full range of construction applications; therefore, sealants are manufactured for interior use, exterior use, gun or pour grade, ability to expand and contract, service temperature range, paintability and for compatability with the material to be sealed.

An oil base caulking compound has a life expectancy of about 4 or 5 years. Its restrictions list a joint maximum size of ½" wide x ¾" deep and is incapable of withstanding any joint movement without rupture. Therefore, using this material for exterior applications may not be prudent or cost effective.

Acrylic latex caulk is probably a good starting point for exterior use. It is considered to be the best value among sealants in the middle performance range due to; relatively inexpensive, life expectancy of 8 to 12 years, cures quickly, excellent paintability, non-staining, and has reasonable elongation and recovery for movement of joints. However, this material has a maximum joint size of ½" wide and ½" deep and should not be used for expansion joints.

Polysulfide base sealants, available in one or two part compounds, are among the premiere compounds on the market. It is reasonably expensive and requires a little more labor time to install, however, its movement capabilities, bonding strength, wide use range (including expansion joints) and a life expectancy of 15± years makes it cost effective while providing excellent joint sealing characteristics.

It is important to have a properly designed joint, both in width and depth. Good sealants will expand and contract with the movement of the joint. As the joint moves, the sealant changes shape while the volume remains constant. It is critical that the width-to-depth ratio be designed to withstand the constant elongation and compression cycles over long periods of time.

Deep beads of sealant wastes material and are more prone to failure than shallow sealant beads. Another important factor to consider is that sealant works best when adhering to the two opposing faces and isolated from the third (back) side of the joint to be sealed. This is accomplished by filling the joint with a material such as oakum or polyethylene foam rod to within ⅜" or ½" of the face of the joint. In shallow depth joints, use a non-adhering tape at the back of the joint to prevent the sealant from bonding.

Sealants are manufactured in colors from white to black and some are clear (silicones). This is necessary to be architecturally compatible with the adjoining surfaces, in addition paint does not adhere at all to some sealants.

The following chart will provide a guide for the amount of sealant required to fill various size joints; expressed in linear feet of joint obtained per gallon of material.

	WIDTH, INCHES							
	1/16	1/8	1/4	3/8	1/2	5/8	3/4	1
1/16	4928	2464	1232	821	616	493	411	307
1/8	—	1232	616	411	307	246	205	154
3/16	—	—	411	275	205	164	137	103
1/4	—	—	307	205	154	123	103	77
3/8	—	—	—	137	103	82	68	51

DEPTH, INCHES

In using the above chart, if the size of the joint to be sealed is not given on the drawings; use the principle that the depth of the sealant is ½ of the width of the joint to be sealed.

When estimating the labor required to caulk a joint, it will depend upon the width of the joint, the material used, and the location or accessibility of the joint. One man should caulk about 600 lin. ft. per 8 hr. day of joints around doors or window frames when they are close to the ground or accessible without the use of scaffolding. This time may be reduced in half if working off ladders or scaffolding. This same man, working on sealing wide expansion joints, will only finish about 200 lin. ft. per 8 hr. day.

CHAPTER 8

DOORS AND WINDOWS

CSI   DIVISION 8

08100   METAL DOORS AND FRAMES

Each hollow metal manufacturer produces their stock design doors and frames as well as a custom line to suit the project Architect's requirements. Bear in mind that the "stock" line will vary between manufacturers, e.g., sizes, width of frames, hardware preparation, etc. For example, a hollow metal door from one manufacturer usually can not be hung in the hollow metal frame from another manufacturer, due to unmatched hinge or lockset locations.

"Stock" will also refer to the cut-out preparation for hinges and, in particular, locksets. Most manufacturers prepare the frames for 3 hinge locations each sized 3½" or 4½" in height and for the standard cylindrical lockset. If the finish hardware specified for the project varies from the manufacturers' standards, the hollow metal manufacturer will consider the order "custom" and increase the price accordingly as well as increase the delivery time to the project.

Hollow Metal Frames.—Frames are made of 18, 16, and 14 gauge metal and formed to receive a 1⅜" or 1¾" thick door, depending on which side of the stop the hinge and strike cut-outs are placed. See the following section through a standard hollow metal frame.

TYPICAL FRAME PROFILE·

Therefore, when the manufacturer varies the overall width of the frame to suit the different wall conditions that may be encountered, he simply increases the width of the stop. This allows them to set up any width frame for either thickness of door in lieu of manufacturing frames for 1⅜" doors and 1¾" doors.

As stated, frames are made in various widths (profile) to suit the thicknesses of the walls to which they are mounted. The standard widths with most manufacturers are 4¾", 5¾", 6¾" and 8¾".

In addition to the frame width dimensions, frames are ordered according to the door opening size, the standard being 2'-0", 2'-4", 2'-6", 2'-8", 3'-0", 3'-4", 3'-6", 3'-8" and 4'-0" wide for single swing and 4'-0", 4'-8", 5'-0", 5'-4", 6'-0", 6'-8", 7'-0", 7'-4" and 8'-0" wide for double swing by 6'-8", 7'-0", 7'-2" and 8'-0" in height.

Frames are available from the factory in welded joint construction (all 3 sides put together to form a unit) or in knocked down (K.D.) form for use on drywall partitions. The K.D. frame is considered to be a pressure-fit frame that once in place (capping the wall) it exerts a gripping action onto the wall. The welded frame is stronger and must be erected prior to the wall erection, whereas, the K.D. frame is erected after the wall is built.

Special frames can be ordered which will incorporate a transom panel above the door and/or sidelites of various heights and widths beside the door.

Average material prices for hollow metal frames of various sizes are as follows:

18 Ga. Welded Hollow Metal Frames

Size	4¾"	5¾"	6¾"	8¾"
2'-6"x7'-0"	$57.00	$59.00	$61.00	$63.00
3'-0"x7'-0"	58.00	60.00	62.00	64.00
5'-0"x7'-0"	73.00	75.00	77.00	79.00
6'-0"x7'-0"	75.00	77.00	79.00	81.00

Note: Deduct $10.00 each for knocked down (K.D.) frames.
Add $10.00 for "B" Label, $15.00 for "A" Label.

Labor Erecting Hollow Metal Frames.—Two carpenters should erect about 16 welded frames per 8 hour day at the following labor cost per frame:

	Hours	Rate	Total	Rate	Total
Carpenter	1.0	$....	$....	$16.47	$16.47

One carpenter should erect about 12 K.D. frames per 8 hour day at the following labor cost per frame:

	Hours	Rate	Total	Rate	Total
Carpenter	0.67	$....	$....	$16.47	$11.04

Hollow Metal Doors.—Doors are made with 20, 18, or 16 gauge face sheets and in 1⅜ " and 1¾ " thicknesses. Door sizes will correspond to the frame opening sizes stated previously in the Hollow Metal Frame data.

Doors may be ordered in a variety of styles including: small vision lites, half glass, full glass, dutch, louvered, and flush. Actually, when a door is ordered with a vision lite, what is received from the manufacturer is a flush door that has been cut out and prepared for future glass installation at the jobsite; the glass is not supplied by the door manufacturer.

The physical construction of the door itself is where the manufacturers really differ. They each have their own techniques for assembly i.e., location of seams, inverted channel or flush top and bottom edges, gauge of hinge and closer reinforcements, leading edge reinforcement, core materials, etc. The contractor is obligated to determine the compliance of the manufacturers product with the requirements contained in the specification. Therefore, be sure of compliance with specification requirements when pricing hollow metal doors, because prices vary widely.

Average material prices for hollow metal doors are as follows:

20 Ga. Hollow Metal Doors

Size	1⅜ "	1¾ "
2'-6"x7'-0"	$125.00	$135.00
3'-0"x7'-0"	135.00	145.00
3'-6"x7'-0"	150.00	160.00

Note: Add $20.00 for "B" Label, $45.00 for "A" Label
Add $60.00 for louver, $50.00 for lite cut-out
Add $40.00 for mineral core (for temperature rise rating)

Labor Erecting Hollow Metal Doors.—Two carpenters should hang, on hinges, about 12 doors per 8 hour day at the following labor cost per door:

	Hours	Rate	Total	Rate	Total
Carpenter	1.34	$ ...	$	$16.47	$22.07
Laborer (for distribution)	.5	....	$	$12.54	6.27
			$		$28.34

Fire Label Requirements.—According to building codes, certain door locations within a building; such as at stairs, trash rooms, furnace rooms, garage entries, storage rooms, etc., will require a fire rating for the door and frame. This rating is expressed as "C Label" for ¾ hour rating, "B Label" for 1½ - 2 hour rating, and "A Label" for 3 hour rating.

When a door is to be fire rated, it also carries other restrictions such as; (1) the latch or lockset must be installed with a UL throw bolt, (2) glass is either not allowed (as in "A Label" rating) or restricted in size (as in "B or C Label" ratings), (3) if

Setting Exterior Wood Door Frames

Size of Door Opening	No. Frames set 8-Hr. Day	Carp. Hrs. per Frame	Add Hrs. for Transom
3'-0"x7'-0"	7-9	0.9-1.1	0.3
6'-0"x7'-0"	5-6	1.3-1.6	0.3
8'-0"x8'-0"	5-6	1.3-1.6	0.3
First Grade Workmanship			
3'-0"x7'-0"	5-7	1.1-1.6	0.3
6'-0"x7'-0"	4-5	1.6-2.0	0.3
8'-0"x8'-0"	4-5	1.6-2.0	0.3

Interior Door Trim

The labor cost of placing interior door trim will vary with the type of trim or casings used and the class of workmanship.

Setting Interior Wood Door Jambs

Size of Door Opening	Description of Jamb	No. Set 8-Hr. Day	Carp. Hrs. per Jamb	Add for Transom
3'-0"x7'-0"	Plain door jambs	8-10	0.8-1.0	0.6
3'-0"x7'-0"	Paneled door jambs	5-6	1.3-1.6	0.6
7'-0"x7'-0"	Plain door jambs	5-6	1.3-1.6	0.6
	First Grade Workmanship			
3'-0"x7'-0"	Plain door jambs	7-9	0.9-1.1	0.6
3'-0"x7'-0"	Paneled door jambs	4-5	1.6-2.0	0.6
7'-0"x7'-0"	Plain door jambs	4-5	1.6-2.0	0.6

a louver is needed in the door, it must have a fusible link that will melt and allow the louver blades to close when the fire reaches the door, and (4) the door must have a closer device or spring hinges.

Fitting and Hanging Wood Doors.—The labor cost of fitting and hanging doors will vary with the class of workmanship, weight of door, whether soft or hard wood, and upon the tools and equipment used. There are power planes, hinge butt routers and lock mortisers that enable a carpenter to prepare about 3 times as much work per day as where the doors are fitted by hand. This must be considered when making up the estimate.

Where just an ordinary grade of workmanship is required, the doors do not always show the same margin at the top and sides, the screws are frequently driven instead of placed with a screw driver and in many instances the doors are "hinge-bound" so they will not open and close freely.

Ordinary Workmanship.—Where doors are fit and hung by hand, a carpenter should fit and hang 8 doors per 8-hr. day, at the following labor cost per door :

	Hours	Rate	Total	Rate	Total
Carpenter	1.0	$....	$....	$16.47	$16.47

On large production line jobs where power tools can be used to advantage and where the same class of work is performed using a power plane, an electric hinge butt router for door and jambs and an electric lock mortiser, a carpenter should fit and hang 12 to 16 doors per 8-hr. day, at the following labor cost per door :

	Hours	Rate	Total	Rate	Total
Carpenter	0.6	$....	$....	$16.47	$9.88

Labor Installing Pre-Hung Door Units.—Pre-hung door units are used extensively in the construction of houses and apartments ranging from "economy" grade to middle class.

In the completely assembled package, jambs are assembled, door is hung in place with hinges applied and stops are mitered and nailed into place.

Where assembled pre-hung door units are used, a carpenter should install 16 units per 8-hr. day, at the following labor cost per unit :

	Hours	Rate	Total	Rate	Total
Carpenter	0.5	$....	$....	$16.47	$8.24

Fitting and Hanging Hardwood Acoustical Doors.—The labor cost of fitting and hanging hardwood acoustical doors is always figured on a first grade workmanship basis as they must be perfectly installed to insure their efficient operation.

Assuming a carpenter, experienced in hanging acoustical

doors, is employed to do the work, the following production
should be obtained :

Type of Door	Approximate Weight per Sq. Ft.	Carpenter Hours per Opening	Helper Hours per Opening
Hardwood Class 36	4 lbs.	6	3
Hardwood Class 41	7 lbs.	8	4

**Fitting and Hanging Heavy Wood Swinging Doors Up to
4'-0"x8'-0".**—When hanging heavy wood swinging doors, such as
used in garages, stables, mill and factory buildings, etc., where the
doors are 3'-0" to 4'-0" wide and 7'-0" to 8'-0" high, it will require
about 4 to 4½-hrs. time to fit and hang one door or 8 to 9-hrs. per
pair, at the following labor cost per door :

	Hours	Rate	Total	Rate	Total
Carpenter	4.3	$....	$....	$16.47	$70.82

**Fitting and Hanging Heavy Wood Swinging Doors Up to
5'-0"x10'-0".**—To fit and hang extra heavy wood swinging
doors up to 5'-0"x10'-0"x2½" or 3" thick, will require about 4
to 5-hrs. time for 2 carpenters, at the following labor cost per
door :

	Hours	Rate	Total	Rate	Total
Carpenter	9	$....	$....	$16.47	$148.23

**Fitting and Hanging Wood Sliding Doors Up to 4'-0"x8'-
0".**—When fitting and hanging heavy wood sliding doors, such
as used in garages, factories, etc., two carpenters working together should place tracks and hang one door complete in 3 to
3½ hrs. at the following labor cost per door :

	Hours	Rate	Total	Rate	Total
Carpenter	6.5	$....	$....	$16.47	$107.06

To place track and hang a pair of double sliding doors for
and opening up to 8'-0"x8'-0", will require 10 to 12 hrs. carpenter time, at the following labor cost per opening :

	Hours	Rate	Total	Rate	Total
Carpenter	11	$....	$....	$16.47	$181.17

The above does not include cutting or drilling in masonry
walls to place bolts or anchors but is based on this work being
done when the walls are built.

**Fitting and Hanging Heavy Wood Sliding Doors Up to
10'-0"x10'-0".**—To place track, fit and hang one sliding door
up to 10'-0"x10'-0" in size and 2½" to 3" thick, will require
about 12 hrs. carpenter time, at the following labor cost per
opening :

	Hours	Rate	Total	Rate	Total
Carpenter	12	$....	$....	$16.47	$197.64

The above does not include cutting or drilling masonry walls for bolts or anchors.

Fitting and Hanging Heavy Wood Sliding Doors Up to 12'-0"x18'-0".—To place track and hang one large door for an opening up to 12'-0"x18'-0", will require about 24 to 28 hrs. carpenter time, at the following labor cost per door:

	Hours	Rate	Total	Rate	Total
Carpenter	26	$....	$....	$16.47	$428.22

It will require 4 to 6 men to handle a door of this size.

Labor on Sliding Door Openings Complete with Pockets.—If pockets are sent to the job with track fitted in place or steel pocket door T-frames are used, and the job carpenters set the box or steel frame and fit and hang the sliding door, it will require about 2 hrs. carpenter time per opening, exclusive of finish hardware, at the following labor cost per opening:

	Hours	Rate	Total	Rate	Total
Carpenter	2	$....	$....	$16.47	$32.94

For double sliding door openings, double the above costs.

Labor Fitting and Hanging Sliding Wood Closet Doors.—When hanging and fitting wood closet doors that slide past each other without the necessity for pockets, a carpenter should place track, attach hangers, fit and hang two doors complete for an opening 3'-0" to 5'-0" wide and 6'-8" to 8'-0" high in about 2 hours.

This does not include time setting door jambs and casing openings on two sides, as this should be added according to time given on previous pages.

The labor cost per opening should average as follows:

	Hours	Rate	Total	Rate	Total
Carpenter	2	$....	$....	$16.47	$32.94

08300 SPECIAL DOORS

TIN CLAD FIRE DOORS

Tin clad fire doors are used in factory, mill and warehouse buildings and may be either sliding or swinging, having fusible links and automatic closing devices.

Sliding doors and lap type swinging doors in brick walls do not ordinarily require a steel frame but they may be necessary with tile walls. Flush type swinging doors require a frame.

Standard tin clad fire doors are usually manufactured in two thicknesses depending on the fire rating required. For Class "A" rating, the thickness is 2⅛" consisting of 3-plies of tongue-and-groove lumber clad with tin sheeting on all sides. For Class

"B" or "C" rating, the thickness is 1¾" consisting of 2-plies of tongue-and-groove lumber with cladding.

Labor Erecting Tin Clad Doors

The following schedules, furnished by the Richmond Fire-proof Door Co., are guides for estimating labor for erection.

The hours of labor are based on brick walls 13" thick. For walls greater in thickness, add ¼ hour for one man for each 4" in thickness per bolt involved. Concrete and stone walls require slightly more time drilling; terra cotta, slightly less. Where "doors both sides" are noted, this assumes doors back to back and tracks bolted together. When tracks cannot be bolted together use "one side" schedule multiplied by two.

		Labor			
		Door One Side		Door Both Sides	
Type	Width	Mech.	Help.	Mech.	Help.
Single	2'-0" to 4'-0"	7	7	10	10
Slide	4'-0" to 5'-2"	8	8	12	12
	5'-3" to 7'-8"	9	9	14	14
	7'-9" to 8'-8"	11	11	17	17
	8'-9" to 12'-0"	12	12	18	18
Slide in	4'-0" to 4'-5"	9	9	14	14
Pairs	4'-6" to 9'-8"	11	11	16	16
	9'-9" to 10'-6"	13	13	18	18
	10'-7" to 12'-0"	15	15	20	20
Vertical	to 4'-10"	13	13	21	21
Sliding	4'-11" to 7'-6"	15	15	24	24
	7'-7" to 10'-2"	17	17	26	26
	10'-3" to 12'-0"	19	19	28	28

		Labor			
		Door One Side		Door Both Sides	
Type	Door Height	Mech.	Help.	Mech.	Help.
Single	to 5'-9"	5	5	7	7
Swing	5'-10" to 8'-9"	5½	5½	8	8
	8'-10" to 12'	6	6	9	9
Swing in	to 5'-9"	8	8	12	12
Pairs	5'-10" to 8'-9"	9	9	14	14
	8'-10" to 12'-0"	10	10	16	16

08330 COILING DOORS

Coiling doors are usually called rolling shutters and are fully factory assembled units made of steel or aluminum. Often, they are labeled fire doors and installation is governed by Underwriter's specifications. Steel mounting jambs and heads are usually furnished by the metal fabrications contractor.

Costs for a manually operated, 20 gauge, door up to 12'x12' will cost around $5.50 to $6.50 a sq. ft., including hood, guides and operating hardware.

Where class A fire ratings are required, doors will cost $5.00 a square foot more. 18 gauge construction will add another $0.80 a sq. ft. Motor operation will cost $500.00 for smaller sizes $600 for medium. Each pass door will add another $500.00.

Two mechanics should hang and adjust a rolling shutter in size up to 12'x12' in one 8 hour day, at the following cost per shutter:

	Hours	Rate	Total	Rate	Total
2 Mechanics	16.0	$....	$....	$17.70	$283.20

Coiling Grilles

Rolling grilles are especially popular now for protection of store fronts, for open fronted stores in enclosed shopping areas and for closing off counter areas.

Stock commercial grilles will cost around $12.00 per sq. ft. with all hardware. As these are often part of a decorative front, they are often made of anodized aluminum, and cost around $15.00 a sq. ft. Exact prices can be obtained from the manufacturer who will also furnish shop drawings upon request to show all the necessary anchoring involved. Steel grilles run around $6.50 a sq. ft. Motorizing and installation costs are the same as for rolling shutters.

08350 FOLDING DOORS

Steel bi-fold closet doors may be installed in openings with drywall, wood, or steel jambs and head frame. Each bi-fold door package comes complete with door panels, track, guides, and pulls for a complete installation.

Doors are available in a variety of styles including flush, panel, half louver, full louver, etc. The doors are shipped "factory finished" with a baked-on paint finish.

A two panel unit is used for openings up to 3'-0" wide and will cost about $45.00 each. A four panel unit is used for openings from 4'-0" to 6'-0" wide, with the 6'-0" unit costing about $90.00 each. Units are sold in 6'-8" and 8'-0" heights, with the 8'-0" heights costing an additional $10-$20 more per unit.

One carpenter should install a complete bi-fold unit and all hardware in about 45 minutes at the following labor cost per unit:

	Hours	Rate	Total	Rate	Total
Carpenter	0.75	$....	$....	$16.47	$12.35

Folding doors, which are often folding partitions, can be ordered completely furnished and with all hardware in wood or vinyl. Folding doors for a 3'x6'-8" opening will run around $60.00 in vinyl, $90.00 in wood slats. Doors for 6'x6'-8" openings will run $140 for vinyl, $170 for wood. Deluxe and custom models will run somewhat more. All the above prices include hardware but the frame must be furnished by others. When the opening is over 150 sq. ft. the units are considered "partitions" rather than doors and will run around $4.25 a sq. ft. for light models to $6.50 a sq. ft. for heavy duty units.

08360 OVERHEAD DOORS

Overhead Garage Doors.—Overhead garage doors are furnished in several different styles, including the one-piece overhead door; self-balancing overhead door and the spring balanced over-head doors.

For an average opening up to 8'-0"x7'-0", two carpenters should place tracks, fit and hang doors and apply the necessary hardware in 4 hours, at the following labor cost per door :

	Hours	Rate	Total	Rate	Total
Carpenter	8	$....	$....	16.47	$131.76

For larger openings up to 16'-0" wide and 7'-0" high, where one wide overhead folding door is used, two carpenters should place tracks, fit and hang doors and apply the necessary hardware complete in 5 hours, at the following labor cost per door :

	Hours	Rate	Total	Rate	Total
Carpenter	10	$....	$....	$16.47	$164.70

Standard Size, Sectional Overhead Type, Garage Doors

Furnished complete with hardware, including cylinder lock, springs and track. Doors may be furnished with all wood panels or top section open for glass. Glass not included. Approximate prices :

Size Opening	1⅜"\nThick	1¾"\nThick
9'-0"x 7'-0"	$225.00	$250.00
14'-0"x 7'-0"	475.00	500.00
15'-0"x 7'-0"	500.00	550.00
16'-0"x 7'-0"	525.00	600.00
10'-0"x10'-0"	....	400.00
12'-0"x10'-0"	....	480.00
10'-0"x12'-0"	....	490.00
12'-0"x12'-0"	....	575.00

Electronic Door Operators

For use where a transmitter is located in each automobile using the garage. A push of the button operates the door from the car.

For residential doors up to 16'-0"x8'-0"
approx. price installed $250.00
An experienced mechanic should install an electric door operator in 2 hours after the necessary wiring is in place.

08370 SLIDING GLASS DOORS

Frame, glazing bead and door members are heavy aluminum
extrusions. Door sections move upon bottom mounted, ball
bearing, grooved brass rollers. Frames are designed with provisions for installing bottom roller, horizontally sliding type
screens. Weatherstripping is mohair pile, factory mounted in
channels to form a continuous double seal. Hardware includes
full-grip lucite pulls and touch latch. Cylinder lock sets are
available at extra cost.

Material Prices of Aluminum Sliding Doors

Type	Frame Size	Economy Crystal Gl.	Deluxe Insulating Gl.
XO	6'-0"x7'-0"	$250.00	$500.00
OXO	9'-0"x7'-0"	300.00	600.00
OXXO	12'-0"x7'-0"	400.00	850.00

For cylinder lock set in lieu of latch, add $10.00.

Labor Erecting Aluminum Sliding Glass Doors.—
Assembly and erection of aluminum sliding glass doors is simple and does not require special tools. All aluminum members
are prefit, with holes drilled and connecting brackets attached.
The following table gives average labor hours required for installing aluminum sliding glass doors in ordinary residential
openings.

Type	Number of Doors	Number of Fixed Panels	Mechanic Hours Required for Assembly	for Erection
XO	1	1	2	6
OXO	1	2	3	8
OXXO	2	2	4	10

Wood Sliding Or Patio Doors

Sliding doors are now available from most of the nationally
distributed millwork houses, and follow the general designs established for aluminum sliding units. These doors can usually
be purchased in two basic widths per panel and in various ar-
rangements including two panels, one fixed, one operating;
three panels, two sides fixed and center panel operable; and
four panels, two end panels fixed and two center panels
operable.

The following price schedule is based on units preglazed with

⅝" safety glass and complete with operating hardware, weatherstripping, sill, and screens for operable sash.

Glass Size	Nominal Door Size	Type	Material Cost
28" wide	5'-0"x6'-8"	Double XO	$ 470.00
34" wide	6'-0"x6'-8"	Double XO	508.00
46" wide	8'-0"x6'-8"	Double XO	668.00
34" wide	9'-0"x6'-8"	Triple OXO	800.00
46" wide	12'-0"x6'-8"	Triple OXXO	1,010.00

Add for keyed lock, $10.00

Grilles which may be removed for cleaning are available in either diamond or rectangular patterns. Costs are $41.00 each for the 28" width, $42.00 each for the 34" width, $50.00 for the 46" width in the rectangular pattern; and $42.50, $44.00 and $62.00 each for the diamond patterns.

08380   SOUND RETARDANT DOORS

Hardwood Sound Insulating Doors

Acoustical doors are for use in consultation rooms, conference rooms, hospitals, clinics, music rooms, television and radio studios, and other areas where acoustical levels must be closely maintained.

Dependent on the desired decibel reduction, or acoustical efficiency, doors are identified and furnished in several models.

Units 1¾" thick are rated in sound transmission class 36. Units 2½" thick are rated in sound transmission class 41. A door 3" thick is also made which has superior acoustical qualities.

Acoustical doors are made in a wide selection of face woods. Custom finishing of doors is available. These are veneered doors and, as such, can be furnished with veneers to match those of the balance of the doors in the building. Any standard hardware may be used for door erection.

The following prices include bottom closers, stops, gaskets and stop adjusters and are for standard rotary cut veneers of natural birch and red oak.

Material Prices of Hardwood Sound Insulating Doors*

Size	1¾" 35 Decibel	2½" 40 Decibel	3" 50 Decibel
2'-6"x6'-8"	$250.00	$325.00	$430.00
2'-8"x6'-8"	260.00	335.00	440.00
3'-0"x6'-8"	270.00	345.00	450.00
2'-6"x7'-0"	260.00	335.00	440.00
2'-8"x7'-0"	270.00	345.00	450.00
3'-0"x7'-0"	280.00	355.00	460.00

*f.o.b. factory, uncrated.

08400   ENTRANCES AND STORE FRONTS

METAL STORE FRONT CONSTRUCTION

Aluminum storefront and entrance doors are manufactured by a number of firms, in a wide variety of styles, to suit the architectural aesthetics desired or a particular design function for most any type of building.

The glass and glazing subcontractor is the firm to contact for this type of work, and they typically represent more than one manufacturer of these materials. This subcontractor will handle all the necessary functions of bidding, shop drawing preparation, ordering, and erection of the storefront materials, as well as the glass, for a complete installation.

Materials are made from an aluminum alloy and extruded into many different shapes and sizes of various thicknesses. The extrusions are offered in stock lengths of 24' long, however, longer lengths are available upon special order.

Installed Costs Of Storefront Construction.—Since there are so many different sizes and shapes of extrusions available for the installation of this type of construction, it would be inappropriate to try to list the material prices and labor time required for the installation, as the labor hours are directly related to the size and shapes of the extrusions to be installed. For instance, a simple storefront and entrance door installation, 40'-0" wide x 10'-0" high, in one system might take 72 man hours, while a more sophisticated system would require 100 man hours.

For budget purposes only, the following installed costs of aluminum storefront framing (glass not included) are given on a per square foot basis for the opening to be enclosed:

1). Windows without intermediate vertical mullions up to 5'-0"x5'-0" in size, using 1¾"x4½" tube - $7.50 per sq. ft.
2). Same as above, but with one vertical mullion - $8.50 per sq. ft.
3). Windows 40'-0"x10'-0" in size with vertical mullions spaced 5'-0" on center, using 1¾"x4½" tube - $6.00 per sq. ft.
4). Same as above, but with 3"x6" tube - $9.50 per sq. ft.
5). For a narrow stile single door and frame, 3'-0"x7'-0" - $550.00 each.
6). For a pair of narrow stile doors and frame, 6'-0"x7'-0" - $850.00 each.

Note: Add 15% for bronze anodized finish
Add 20% for black anodized finish
Add for transom over single door, $80.00; over pair of doors, $140.00

08500   METAL WINDOWS

Labor Installing Steel Windows and Doors.—Installation costs will vary with the class of labor used, type and size of individual units and job working conditions. However, a rea-

sonably accurate estimate of installation labor costs can be
made by multiplying the number of man hours required to install each unit by the prevailing hourly rate of pay, plus rates
for pension, welfare, etc., in effect in the area in which job is
located.

Product	Window Opening	Man Hours Each Unit
Pivoted, Commercial Projected and Architectural Projected Windows	1. Single Unit*	2.25
	2. Two or more Units*	
	(a) Precast Sills	1.25
	(b) Poured Sills	1.50
Fixed Industrial	1. Single Unit	1.75
	2. Two or More Units	
	(a) Precast Sills	.75
	(b) Poured Sills	1.00
Fixed Architectural	1. Single Unit	2.00
	2. Two or More Units	
	(a) Precast Sills	1.00
	(b) Poured Sills	1.25
Intermediate Windows (Casement, Combination and Projected)	1. In Wood	
	(a) Single Unit	2.75
	(b) Two or More Units	2.25
	2. In Stone	
	(a) Single Unit	3.00
	(b) Two or More Units	2.50
	3. In Masonry	
	(a) Single Unit	2.25
	(b) Two or More Units	1.25
Mullions Only		.50
Mullions with Covers		.75
Residence Casements (Roto)	1. Single Unit	
	(a) In Wood	1.75
	(b) In Masonry or Brick Veneer	1.25
	2. Two or More Units	
	(a) In Wood	1.50
	(b) In Masonry or Brick Veneer	1.00
Residence Casements (Fixed)	1. Single Unit	
	(a) In Wood	1.50

	(b) In Masonry or Brick Veneer		1.00
	2. Two or More Units		
	(a) In Wood		1.25
	(b) In Masonry or Brick Veneer		.75
Casements in Casings	Single or Multiple Units		1.50
Continuous Top Hung, Fixed	Per Lineal Foot		.50
Continuous Top Hung, Swing	Per Lineal Foot		.75
Mechanical Operators, Rack and Pinion	Per Lineal Foot		.75
Mechanical Operator, Lever Arm	Per Lineal Foot		.50
Mechanical Operators, Tension	Per Lineal Foot		1.50
Industrial Doors and Frames	1. Up to 35 Sq. Ft. Per Sq. Ft.		.125
	2. Over 35 Sq. Ft. Per Sq. Ft.		.10
Double Hung Windows	1. Single Units		1.50
	2. Two or More Units		1.25
Screen Installation For Projected Windows	Net per Screen		1.10
For Basement and Utility Windows	Net per Screen		.75
For Security Windows	Net per Screen		1.10
For Casement Windows	Net per Screen		.75
For Pivoted Windows	Net per Vent		1.50

*Single ventilator units. For each additional ventilator, ADD........ .25

08600   WOOD AND PLASTIC WINDOWS

Wood windows are most generally sold as complete units with frame, sash, operating hardware, weatherstripping and glazing assembled at the factory.

Setting Single Wood Window Frames.—The labor cost of setting wood window frames will vary with the size of the frame, amount of bracing necessary and the distance they have to be carried or hoisted.

After a frame has been delivered at the job, it will require 10 to 15 minutes sorting frames and carrying them to the building and it will then require 20 to 30 minutes for 2 carpenters to set, plumb and brace each frame, which is at the rate of 8 to 12 frames per 8-hr. day.

The labor handling and setting each frame should cost as follows:

	Hours	Rate	Total	Rate	Total
Carpenter	0.8	$....	$....	$16.47	$13.18
Labor	0.2	....	....	12.54	2.51
Cost per frame ..			$....		$15.69

Labor Handling and Setting Complete Window Units.—When handling and setting complete window units consisting of frame, sash, balances, weatherstrips, etc., all installed complete, figure about ¼-hr. more than for handling and setting window frames of the same size.

Fitting and Hanging Wood Sash.—A carpenter should fit and hang, complete with sash cord, chain and weights or counterbalances, 12 to 14 single sash per 8-hr. day, at the following labor cost per pair:

	Hours	Rate	Total	Rate	Total
Carpenter	1.2	$....	$....	$16.47	$19.76

Fitting and Hanging Casement Sash.—Under average conditions, a carpenter should fit and hang 10 to 13 single casement sash per 8-hr. day, at the following labor cost per pair:

	Hours	Rate	Total	Rate	Total
Carpenter	1.4	$....	$....	$16.47	$23.06

Fitting and Hanging Outside Window Shutters.—When fitting and hanging outside window shutters on medium priced residences, cottages, etc., a carpenter should fit and hang 7 to 8 pairs per 8-hr. day, at the following labor cost per pair:

	Hours	Rate	Total	Rate	Total
Carpenter	1.1	$....	$....	$16.47	$18.12

Interior Window Trim

The labor cost of placing interior window trim will vary considerably with the type of building, thickness of walls, style of trim or casing, etc. Frame buildings, stucco buildings, brick veneer buildings, etc., having walls 6" to 8" thick will not require inside wood jamb linings, as the wood window frames usually extend to the inside face of the wall and the wood casings, window stools, aprons, stops, etc., are nailed directly to the wood frame.

Buildings having walls 10", 12", 16" or more in thickness will require a wood jamb lining to extend from the inside of the wood window frame to the face of the finished wall. It will also require wood blocking and grounds to nail the jamb linings, window stools, aprons, etc.

The style of trim will affect labor costs considerably and for that reason costs are given on 4 distinct types of interior casings, as follows:

Style "A" consists of 1-member trim with either a mitered or square cut head or "cap". This is the simplest trim it is possible to use.

Style "B" consists of 1-member trim for sides and a built-up cap trim for window heads. Caps may be put together in the mill or assembled and put together on the job.

Style "C" consists of 2-member back-band trim, consisting of a one-piece casing with a back-band of thicker material.

Style "D" consists of any type of window trim that are assembled and glued together in the mill and set in the building as a unit.

Labor Time Erecting Interior Window Trim

Style Trim	Kind of Trim	No. Windows 8-Hr. Day	Carp. Hrs. per Window	Add for Wood Jamb Lining
"A"	Single Casing	7–8	1.0–1.1	0.6
"B"	Cap Trim*	5–7	1.1–1.6	0.6
"C"	Back-bank Trim	5–6	1.3–1.6	0.6
"D"	Mill Assembled**	8–10	0.8–1.0	0.6
	First Grade Workmanship			
"A"	Single Casing	5–7	1.1–1.6	0.8
"B"	Cap Trim*	4–6	1.3–2.0*	0.8
"C"	Back-band Trim	4–5	1.6–2.0	0.8
"D"	Mill Assembled**	7–9	0.9–1.1	0.8

*For cap trim put together on job.
**Trim assembled and glued up in the mill.

08700 HARDWARE AND SPECIALTIES

08710 FINISH HARDWARE

Finish hardware refers to any item that is usually fitted to a door and frame to perform a specific function, e.g., hinges, latch or locksets, closers, stops, bolts, etc.

Most items of finish hardware can be obtained in various finishes and each finish has a U.S. Code Symbol Designation as follows:

US P	-	Prime Paint Coat
US 3	-	Polished Brass
US 4	-	Satin (dull) Brass
US 9	-	Polished Bronze
US 10	-	Satin (dull) Bronze
US 10B	-	Satin Bronze - Oil Rubbed
US 14	-	Polished Nickel
US 15	-	Satin (dull) Nickel
US 20	-	Statuary Bronze

US 26 - Polished Chrome
US 26D - Satin (dull) Chrome
US 28 - Satin Aluminum - Anodized
US 32 - Polished Stainless Steel
US 32A - Satin (dull) Stainless Steel

The specification writer will refer to a specific Federal Specification Series Designation when specifying the type of lockset required for the project, and each of these series relates to the construction and design function of the lockset. For example:

ANSI* Series 1000 - Mortise type latchset or lockset and keyed deadbolt housed in one casing.
ANSI* Series 2000 - Mortise type latchset or lockset with integral deadbolt operated by the key in the knob.
ANSI* Series 4000 - Cylindrical latch and locksets
        Grade #1 - Heavy Duty
        Grade #2 - Medium Duty
        Grade #3 - Light Duty

*American National Standards Institute

The following listing of finish hardware material prices should be used as a guide only; always obtain a quotation from a material supplier for the specific type, function, finish, and style of finish hardware specified for the project.

Material Prices For Finish Hardware

Item	Price Range
Cylindrical Locksets	
Std. Duty	$ 15.00-$ 45.00
Heavy Duty	50.00- 80.00
Mortise Locksets	
Heavy Duty	55.00- 95.00
Push-Pull Bars	50.00- 150.00
Anti-Panic Device, Single	150.00- 250.00
Hinges (per pair)	
Steel Plain	10.00- 15.00
Steel B.B.	25.00- 50.00
Bronze B.B.	75.00- 100.00
Door Closers	
III	45.00- 75.00
IV	55.00- 80.00
V	65.00- 100.00
Surface Bolt	6.00- 15.00
Kick Plate: Alum	5.00- 15.00
Bronze	15.00- 35.00
Door Stops:	
Floor Bumper	3.50- 4.50
Wall Bumper	7.50- 9.00
Bumper Plus Holder	16.00- 20.00
Door Plunger	15.00- 18.00
Thresholds	
Alum.	4.00- 16.00
Bronze	25.00- 65.00

The following table will give average production time, in carpenter hours, necessary for the installation of finish hardware items:

Items of Finish Hardware	Hours to Install
Cylindrical Locksets (door prepared)	0.5
Mortise Locksets (door prepared)	0.7
Push-Pull Bars	0.4
Anti-Panic Device	1.5
Door Closers	1.0
Surface Bolts	0.3
Flush Bolts	1.0
Door Stops - Wall Type	0.3
Floor Type	0.5
Thresholds	0.5

08800 GLAZING

The new process of making glass consists of floating molten glass over large pools of molten tin. Through automated production stages, the floated glass is fire polished to remove all distortion, allowed to cool, and fed onto the tables in a continuous ribbon where it is cut to sheet size for packing into shipping crates. The glass is classified as "float glass" and is made in glazing quality and mirror quality of various thicknesses.

Clear Float Glass - Glazing Quality

Product	Thickness	Wt./Sq. Ft.	Standard Max. Size	Material Cost/ Sq. Ft.
Float	3/32"	1.22 lbs.	40"x100"	$1.10
	1/8"	1.62	80"x120"	1.25
	3/16"	2.43	120"x212"	1.35*
Float/Plate	1/4"	3.24	130"x212"	1.50*
	3/8"	4.92	124"x204"	2.50*
	1/2"	6.56	124"x204"	3.75
	3/4"	9.85	124"x204"	5.50

*Note: Add $0.75 per sq. ft. for gray or bronze tint

Float glass is also available in glare and heat reducing tints of gray and bronze, heat absorbing (blue-green), tempered, coated with thin metallic coverings for reflective qualities, laminated, and in insulating glass units consisting of two sheets of glass separated by an internal air space.

Labor Setting Window or Plate Glass in Wood Sash, Using Putty or Glazing Compound

Approximate Size of Glass	Number Lights Set 8-Hr. Day	Glazier Hours per 100 Lights
12"x14"	65	12.5
20"x28"	40	20.0
30"x40"	30	26.7
40"x48"	20	40.0

Putty Required Setting Glass in Wood Sash.—Window glass set in 1⅜" wood sash requires 1-lb. of putty for each 8 to 8½ lin. ft. of glass or sash rabbet; when set in 1¾" sash it requires 1-lb. of putty to each 7¼ to 7½ lin. ft. of glass or sash rabbet.

The following table gives the quantity of putty required for glazing various size glass :

Sash Thick	Size of Glass Inches				
	12"x14"	14"x20"	20"x28"	30"x36"	40"x48"
13/8"	1/2-lb.	2/3-lb.	1 lb.	13/8-lb.	13/4-lb.
13/4"	3/5-lb.	3/4-lb.	11/8 lb.	11/2-lb.	2 -lbs.

Labor Glazing Steel Sash.—On jobs where there are no unusual window heights or other abnormal conditions, a glazier should set the following number of lights of glass in the various types of steel sash per 8-hr. day.

In clerestory, monitor or outside setting (from scaffold or swing stage), it will require one man on the ground for each two men on the scaffold to keep them supplied with glass, putty, etc. This adds approximately 50 percent to the costs as given below.

Add up to 25 percent to the following costs for winter work.

Pivoted Steel or Commercial Projected Steel Windows*

Glass Size Inches	Number Lights per 8-Hr. Day	Glazier Hrs. per 100 Lts.	Lbs. Putty per Light
22"x16"	50	26.7	1.25
32"x16"	40	26.7	1.50
32"x22"	30	32.0	1.67

Architectural Projected Steel Windows**

18"x16"	30	23.0	1.25
20"x16"	30	23.0	1.25
38"x16"	25	40.0	1.75
40"x16"	25	50.0	2.00
46"x16"	25	50.0	2.00
48"x16"	25	50.0	2.25

*Based on setting glass from inside and working from floor.
**Based on setting glass from outside, using glazing compound or steel sash putty.

CHAPTER 9

FINISHES

CSI DIVISION 9

09100 LATH AND PLASTER

Estimating Quantities of Lathing and Plastering.—
Lathing and plastering are estimated by the square yard, but
the method of deducting openings, etc., varies with the class of
work and the individual contractor.

Many plastering contractors make no deductions for door or
window openings when estimating new work, while others deduct for one-half of all openings over a certain size.

GYPSUM LATH

Sizes and Weights of Gypsum Lath

Size and Thickness	Wt. Per Sq. Yd.	Approx. Price 1,000 Sq. Ft.
16"x48"x⅜"	14 Lbs.	$105.00
16"x48"x½"	19 Lbs.	$110.00

Labor Cost of 100 Sq. Yds. of ⅜" Gypsum Lath Applied to Wood Studs or Joists with Nails

	Hours	Rate	Total	Rate	Total
Lather	8	$....	$....	$16.35	$130.80
Cost per sq. yd.			....		1.31

Material Cost of 100 Sq. Yds. Suspended Gypsum Lath Ceilings Using 16"x48"x⅜" Gypsum Lath with Main Runners 3'-0" on Centers and Cross Furring Channels 16" on Centers

	Rate	Total	Rate	Total
77 pcs. 3/16"x3'-0" hangers, 231'-0"	$....	$....	$.03	$ 6.93
364 lin. ft. 1 1/2" channels	....	....	.24	87.36
700 lin. ft. 3/4" channels	....	....	.16	112.00
945 sq. ft. 16"x48"x3/8" gypsum lath	....	....	.105	99.23
500 Brace-tite field clips	....	....	.08	40.00
340 Bridjoint clips, B-1	....	....	.05	17.00
Cost 100 sq. yds.		$....		$362.52
Cost per sq. yd.		....		3.63

Labor Cost of 100 Sq. Yds. Suspended Gypsum Lath Ceilings Using 16"x48"x⅜" Gypsum Lath with Main Runners 3'-0"on Centers and Cross Furring Channels 16" on Centers

	Hours	Rate	Total	Rate	Total
Lather	20	$....	$....	$16.35	$327.00
Labor	4	....	....	13.77	55.08
Cost 100 sq. yds.			$....		$382.08
Cost per sq. yd.		....			3.82

METAL LATH AND FURRING

		Maximum Allowable Spacing of Supports in Inches				
	Min. Wt. Lath	Vertical Supports			Horizontal	Supports
Type of Lath	Lbs. per Sq. Yd.	Wood	Metal Solid Partit.	Others	Wood or Concrete	Metal
Flat Expanded	2.5	16	16	12	0	0
Metal Lath	3.4	16	16	16	16	13½
Flat Rib	2.75	16	16	16	16	12
Metal Lath	3.4	19	24	19	19	19
⅜" Rib	3.4	24	..	24	24	24
Metal Lath	4.0	24	..	24	24	24
Sheet Lath*	4.5	24	..	24	24	24
Wire Lath	2.48	16	16	16	13½	13½
V-Stiffened						
Wire Lath	3.3	24	24	24	19	19
Wire Fabric	..	16	0	16	16	16

*Used in studless solid partitions.

Hanger Sizes for Suspended Ceilings

Ceiling Area, SF Max.	Hanger Size Min.	Hangers for Supporting up to 25 SF Ceiling	
8	12 ga. wire	6 ga. wire	1 "x3/16" flat 1 1/4"x 1/8"
12	10 ga. wire	5 ga. wire	flat 1 1/4"x3/16"
12½	9 ga. wire	3/16" rod	flat 1 1/2"x 1/8"

16	8 ga. wire	7/32" rod	flat
			1 1/2"x3/16"
17½	7 ga. wire	1/4" rod	flat

All wire hangers should be galvanized steel, all flat and rod hangers should be coated with rust-inhibitive paint.

Spans and Spacing for Main Runners in Suspended Ceilings

Min. Size and Type	Max. Span Between Hangers or Supports	Max. CtoC Spacing of Runners
¾" channel, .3 lb.	3'-0"	2'-4"*
¾" channel, .3 lb.	2'-6"	2'-6"
¾" channel, .3 lb.	2'-0"	3'-0"
1½" channel, .475 lb.	5'-0"	2'-0"
1½" channel, .475 lb.	4'-0"	3'-0"
1½" channel, .475 lb.	3'-6"	3'-6"
1½" channel, .475 lb.	3'-0"	4'-0"
2 " channel, .59 lb.	7'-0"	2'-0"
2 " channel, .59 lb.	5'-0"	3'-6"
2 " channel, .59 lb.	4'-6"	4'-0"

All weights of channel are for cold-rolled members.
*This spacing for runner channels supporting furring members against concrete joists, only.

Spans and Spacing for Cross Furring

Min. Size and Type	Max. Span Between Runners or Supports	Max. CtoC Spacing of Cross Furring
⅜" pencil rod	2'-6"	12"
⅜" pencil rod	2'-0"	19"
¾" channel, .3 lb.	4'-0"	16"
¾" channel, .3 lb.	3'-6"	19"
¾" channel, .3 lb.	3'-0"	24"

All weights of channel are for cold-rolled members.

Material Cost of 100 Sq. Yds. of 3.4 Lb. Diamond Mesh Metal Lath Applied to Wood Studs or Joists @ 16" O.C., Ready to Receive Plaster

	Rate	Total	Rate	Total
105 sq. yds. diamond mesh	$....	$....	$1.50	$157.50
9 lbs. 1½" roofing nails.	....		1.00	9.00
Cost 100 sq. yds.		$....		$166.50
Cost per sq. yd.		....		1.67

Labor Cost of 100 Sq. Yds. of 3.4 Lb. Diamond Mesh
Metal Lath Applied to Wood Studs or Joists @ 16" O.C.,
Ready to Receive Plaster

	Hours	Rate	Total	Rate	Total
Lather	7.5	$....	$....	$16.35	$122.63
Cost per sq. yd. ...					1.23

Material Cost of 100 Sq. Yds. Suspended Metal Lath
Ceiling Using 3.4 Lb. ⅜" Rib Metal Lath With Furring
Channels 24" on Centers Runner Channels 36" on
Centers, Hangers 48" on Centers

	Rate	Total	Rate	Total
100-9 ga. x 3'-0" hangers 300'-0" or 19 lbs.	$....	$....	$ .70	$ 13.30
345 lin. ft. 1½" c.r. channels	....	....	.24	82.80
490 lin. ft. ¾" c.r. channels	....	....	.16	78.40
3 lbs. 16 ga. wire	....	....	.80	2.40
105 sq. yds. ⅜" rib metal lath	....	....	2.20	231.00
5 lbs. 18 ga. wire	....	....	.90	4.50
Cost per 100 sq. yds. ..	....	$....		$412.40
Cost per sq. yd.		....		4.13

Labor Cost of 100 Sq. Yds. Suspended Metal Lath Ceiling
Using 3.4 Lb. ⅜" Rib Metal Lath With Furring Channels
24" on Centers Runner Channels 36" on Centers, Hangers
48" on Centers

	Hours	Rate	Total	Rate	Total
Lather*	50	$....	$....	$16.35	$817.50
Cost per sq. yd. ...			....		8.18

*For erection under concrete, add for placing hangers in
forms before pour; or, if necessary to drill holes in concrete or
tile, and place inserts.

Material Cost of 100 Lin. Ft. of Metal Corner Bead
Attached to Lath Bases

	Rate	Total	Rate	Total
100 lin. ft. corner bead ..	$....	$....	$.15	$15.00
Staples or tie wire	....	....		1.00
Cost per 100 lin. ft.		$....		$16.00
Cost per lin. ft.		....		.16

Labor Cost of 100 Lin. Ft. of Metal Corner Bead Attached
to Lath Bases

	Hours	Rate	Total	Rate	Total
Lather	2.5	$....	$....	16.35	$40.88
Cost per lin. ft.			....		.41

ESTIMATING PLASTER QUANTITIES

Gypsum plasters are sold by the ton and are usually packed in 100 lb. sacks. When sand is mixed with plaster in the proportion of 1 to 1, 1 to 2, or 1 to 3, it means 100, 200, or 300 lbs. of sand respectively are to be added to each 100 lb. sack of plaster.

Number of 100 Lb. Sacks of Gypsum Cement Plaster
Required Per 100 Sq. Yds.

	Kind of Plastering Surface	
Metal Lath	Gypsum Lath	Unit Masonry
18 to 20	9 to 11	10 to 12
(Sanded 1 :2;1: 3)	(Sanded 1 :2;1 :3)	(Sanded 1 :3)

For each ⅛" in Thickness, Proportions 1 :3; Add or Deduct
Sacks

2.5	2.5	2.5

Recommended Proportions and Quantities

In the following table giving recommended proportions for gypsum plaster and perlite or vermiculite, the proportions 1-2, means 100 lbs. of gypsum plaster to 2 cu. ft. of perlite or vermiculite; 1-3 means 100 lbs. of gypsum plaster to 3 cu. ft. of perlite or vermiculite.

Type of Construction			Coat	Recommended Volume Proportions	Required per 100 Sq. Yds. Bags Plaster	Cu. Ft. Aggregate
Metal Lath	3/4"	1/4"	Scratch	1—2	10	20
		7/16"	Brown	1—2	12½	45
Gypsum Lath	7/8"	3/16"	Scratch	1—2	5	10
		1/4"	Brown	1—2	10	20
		3/8"	Scratch (One side only)	1—2	15	30
2" Solid Partition	2"	3/8"	Brown (One side only)	1—2	15	30
		1 1/8"	Brown 2nd Side	1—2	45	90

Type of Construction		Coat	Recommended Volume Proportions	Required per 100 Sq. Yds. Bags Plaster	Cu. Ft. Aggregate
Unit Masonry	5/8"	3/16" Scratch	1—3	4	12
		3/8" Brown	1—3	8	24

Note: Grounds attached to studs, joists or channels—not the lath. 1/16" allowed for finish coat thickness.

Gypsum Ready-Mix Plasters Made with Lightweight Aggregate

Gypsum ready-mix plasters are generally made with perlite and the regular formula for lath bases is packed in 80-lb. sacks and the masonry formula for application over all masonry bases is packed in 67-lb. sacks.

The following table gives the number of sacks required per 100 sq. yds. applied over various surfaces:

Kind of Plastering Surface

Metal Lath	Gypsum Lath	Unit Masonry
32 to 34*	17 to 19*	24 to 26**

For Each ⅛" in Thickness, Add or Deduct Sacks

| 5 | 5 | 5 |

*80-lb. sacks.   **67-lb. sacks.

Finishing Lime.—Finishing lime is essential to two common types of job mixed finish coats—the white putty coat trowel finish and the sand float finish.

For the trowel finish coat, the proportion is generally 1 part of gypsum gauging plaster to 2 parts of hydrated lime on a dry weight basis, or 100 lbs. of gypsum gauging plaster to 4½ cu. ft. of lime putty. One ton of gypsum gauging plaster and 2 tons of hydrated finishing lime, or 90 cu. ft. of lime putty will finish approximately 1,200 sq. yds.

While a sand float finish may be mixed with 1 part of lime putty and 2½ to 3 parts of sand, it is generally necessary to add Keene's cement or gypsum gauging plaster to give the finish early hardness and strength. A standard mix for a Keene's cement-lime putty sand float finish is 2 parts of lime putty to 1½ parts of Keene's cement and 4½ parts of sand by volume. This is equivalent to 1 ton of hydrated lime, or 45 cu. ft. of putty, to 1¼ tons of Keene's cement and 4¾ tons of sand and this quantity will cover approximately 1,600 sq. yds.

Prepared Gypsum Trowel Finishes.—There are a number of prepared gypsum trowel finishes on the market which require only the addition of water to make them ready for use. A few of these are Universal White Trowel Finish, Red Top* Trowel, and Red Top* Imperial.

Under average conditions, if mixed and applied in accordance with standard specifications, one ton will cover 350 to 400 sq. yds. of surface or it requires 500 to 575 lbs. per 100 sq. yds.

Prepared Gypsum Sand Float Finishes.—There are a number of gypsum sand float finishes, such as Red Top* Sand Float Finish.

These finishes are ready for use when water is added and under average conditions one ton is sufficient for 250 to 275 sq. yds. or it requires 725 to 85 lbs., per 100 sq. yds.

Estimating Labor Costs of Plastering

Applying Scratch Coat of Plaster.—A plasterer should apply 140 to 165 sq. yds. of scratch coat per 8-hr. day, where ordinary grade of workmanship is required, and 135 to 155 sq. yds. per 8-hr. day where first grade workmanship is required.

Applying Brown Coat.—Where an ordinary grade of workmanship is required, a plasterer should apply 90 to 110 sq. yds. of brown coat per 8-hr. day.

Where first grade workmanship is required, a plasterer should brown out 70 to 85 sq. yds. per 8-hr. day.

Applying "Scratch—Double-Back" Basecoat.—Over masonry bases and in many markets over gypsum lath base it is common practice to apply a single plaster mix in a "scratch-in" and "double-back" operation to fill out to grounds.

For ordinary work, a plasterer should apply 90 to 110 sq. yds. per 8-hr. day, and for first grade workmanship, a plasterer should apply 70 to 85 sq. yds. per 8-hr. day in this manner.

Applying White Finish or "Putty" Coat.—On an ordinary job of white coating, a plasterer should apply 90 to 110 sq. yds. per 8-hr. day.

Where first grade workmanship is specified, a plasterer will apply 60 to 80 sq. yds. per 8-hr. day.

Labor Applying 100 Sq. Yds. of Plaster to Various
Plaster Bases Ordinary Workmanship

Description of Work	Sq. Yds. per 8-Hour Day	Plasterer Hours	Tender Hours
Gypsum Mortar on Gypsum lath—½" Plaster— 2-Coat Work—			
Brown Coat 1 :2½	90–110	8	5
Gypsum Mortar on Unit Masonry—⅝" Plaster— 2-Coat Work—			
Brown Coat 1 :3.................	80–100	9	6

Description of Work	Sq. Yds. per 8-Hour Day	Plasterer Hours	Tender Hours
Lime Mortar on Unit Masonry—⅝" Plaster— 2-Coat Work			
Brown Coat 1 :3	90–100	8.5	5.5
Lime Mortar on Metal Lath—¾" Grounds—⅝" Plaster— 3-Coat Work			
Scratch Coat 1 :2	150–160	5	5
Brown Coat 1 :3	90–110	8	5
Lime Mortar on Unit Masonry—⅝" Plaster— 3-Coat Work			
Scratch Coat 1 :2	150–160	5	5
Brown Coat 1 :3	90–110	8	5
Gypsum Mortar on Unit Masonry Tile—⅝" Plaster— 3-Coat Work			
Scratch Coat 1 :3	150–160	5	5
Brown Coat 1 :3	90–110	8	5
Gypsum Mortar on Gypsum Lath—½" Plaster— 3-Coat Work			
Scratch Coat 1 :2	150–160	5	5
Brown Coat 1 :3	90–110	8	5
Gypsum Mortar on Metal Lath—¾" Grounds—⅝" Plaster— 3-Coat Work			
Scratch Coat 1 :2	150–160	5	5
Brown Coat 1 :3	85–100	9	6
Two (2") Inch Solid Plaster Partitions—Metal Lath One Side—Back Up—Finished 2 Sides—Gypsum Mortar—¾" Grounds—2" Plaster			
Scratch Coat 1 :2	150–160	5	5
Back Up Coat 1 :2	150–160	5	5
2 Brown Coats 1 :2	40– 50	18	12

Labor Applying 100 Sq. Yds. of Plaster to Various Plaster Bases Ordinary Workmanship

Description of Work	Sq. Yds. per 8-Hour Day	Plasterer Hours	Tender Hours
Hollow Stud Partitions—Lathed 2 Sides—Plastered to ¾" Grounds			

Scratch Coat 1 :2—2			
Sides	75– 80	10	10
Brown Coat—2 Sides	40– 50	18	12
Finish White Coat—All			
Classes of Undercoat	90–110	8	4
Sand Finish Coat—All			
Classes of Undercoat	100–120	7.25	5
First Grade Workmanship—Where First Grade Workmanship is Required as described on the previous pages, add to Hours Given above			
Brown Coat	....	2.25	1
Finish White Coat	....	3.50	1
Sand Finish	....	2.75	1

Labor Applying 100 Sq. Yds. Portland Cement Plaster to Various Plaster Bases

Description of Work Ordinary Work*	No. Sq. Yds. per 8-Hr. Day	Hrs. 100 Sq. Yds. Plasterer	Tender
Interior Work on Brick, Clay Tile or			
Cement Block—³⁄₄" Thick			
Scratch Coat	120-140	6	4
Brown Coat	75- 85	10	5
Float Finish Coat	60- 70	12	4
Trowel Finish Coat	45- 55	16	4
Interior Portland Cement Plaster on Metal Lath			
Scratch Coat	120-140	6	5
Brown Coat	60- 70	12	6
Float Finish Coat	60- 70	12	4
Trowel Finish Coat	45- 55	16	4
Exterior Portland Cement Stucco on Brick, Clay Tile or Cement Block			
Scratch Coat	120-140	6	5
Brown Coat	75- 85	10	6
Scaffold	....	..	12
Exterior Portland Cement Stucco on Metal Lath on Frame Construction			
Scaffold	....	..	12
Scratch Coat	120-140	6	6
Brown Coat	60- 70	12	8
Floated Finish Coat	60- 70	12	6
Troweled Finish Coat	45- 55	16	6
Textured Finish Coat	50- 60	14	6

Labor Applying 100 Sq. Yds. Portland Cement Plaster to
Various Plaster Bases (Con't)

Blocking Off Portland Cement Plaster into Squares to Represent Tile	45- 55	16	4
Portland Cement Straight Base	45- 55**	16	4
Portland Cement Coved Base	30- 35**	24	6

*For first grade workmanship, add 30 to 40 percent to plasterer time for brown and finish coats.
**Lineal feet.

EXTERIOR STUCCO

Exterior stucco is generally composed of a portland cement base. The scratch and brown coats should be mixed in the proportions of 1 part of portland cement by weight to 3 parts of clean sand by weight. Do not add more than 8 lbs. of lime to each 100 lbs. of portland cement used in the mixture.

Applying Scratch Coat of Portland Cement Stucco.—Where just an ordinary grade of workmanship is required, a plasterer should apply 120 to 140 sq. yds. of scratch coat per 8-hr. day on metal lath.

Applying Brown Coat of Portland Cement Stucco.—A plasterer should apply 75 to 85 sq. yds. of brown coat per 8-hr. day, on brick, tile or concrete block walls and 60 to 70 sq. yds. per 8-hr. day on metal lath.

Applying Trowel Finish Coat of Portland Cement.—A plasterer should apply 45 to 55 sq. yds. of portland cement trowel finish per 8-hr. day, where just an ordinary grade of workmanship is required and 35 to 40 sq. yds. per 8-hr. day, where first grade workmanship is required.

Applying Wet Rough Cast Finish to Portland Cement Stucco.—If a wet rough cast finish is used and the mortar and aggregate are thrown against the wall with a paddle or similar tool, a plasterer should complete 30 to 40 sq. yds. per 8-hr. day.

Applying Pebble Dash or Dry Rough Cast.—When applying a pebble dash or dry rough cast finish where the aggregate is thrown against the wet cement or "butter" coat, a plasterer should apply 35 to 40 sq. yds. per 8-hr. day.

Washing Exterior Stucco With Acid to Expose Crystals.—Where the finish coat of stucco contains granite, marble or crystal screenings, it is washed off with a solution of muriatic acid to expose the crystals. A man should wash 475 to 525 sq. ft. (53 to 58 sq. yds.) per 8-hr. day.

09250   GYPSUM WALLBOARD

Size and Thickness.—Gypsum wallboard panels are manufactured in a number of thicknesses and sizes with various longitudinal edge designs.

Thickness	Size	Edge Design
¹⁄₄ "	4'-0"x8'-12'-0"	S.E., T.E.
³⁄₈ "	4'-0"x8'-12'-0"	S.E., T.E.
¹⁄₂ "	4'-0"x8'-12'-0"	S.E., T.E.
⁵⁄₈ "	4'-0"x8'-12'-0"	S.E., T.E.

S.E. - Square Edge
T.E. - Tapered Edge

Foil-Back Gypsum Panels.—These panels are made by laminating a sheet of aluminum foil to the back surface of the gypsum wallboard. The foil reduces outward heat flow in winter and inward heat flow in summer, has a significant thermal insulating value if facing an air space of ³⁄₄" or more, and is effective as a vapor barrier.

Moisture Resistant Gypsum Panels.—MR board has a special asphalt composition gypsum core and is covered with chemically treated face papers to prevent moisture penetration. This board is a little harder to cut, has a brownish color core, and is usually covered with a light green finish face paper. These panels were developed for installation in bathrooms, kitchens, utility rooms, and other high moisture areas.

Fire Rated Gypsum Panels.—A specially formulated mineral gypsum core is used to make panels in ¹⁄₂" and ⁵⁄₈" thicknesses for application to walls, ceilings, and columns where a fire rated assembly is needed. Based on tests by Underwriters' Laboratories, Inc., certain wall, floor/ceiling, and column assemblies give 45 minute to 4 hour fire resistance ratings.

Estimating Material Quantities.

Fasteners (nails or screws) should be estimated as requiring about 1,000 each per 1,000 sq. ft. of board to be installed (or 1 fastener per sq. ft.).

It will require about 400 L.F. of joint tape and about 1.3 5-gal. cans of joint compound to finish 1,000 sq. ft. of drywall surface. The amount of compound may be reduced to 1 5-gal. can per 1,000 sq. ft. if there are very few external corners to be finished.

Example: The following list of material quantities may be required for installation and finishing of 300 lin. ft. of drywall partition, 8'-0" high, covering both sides of partition (300 L.F. x 8' Ht. x 2 sides = 4,800 sq. ft.).

"Material Only"

Item	Rate	Total	Rate	Total
¹⁄₂" Drywall Reg. 4,800 sq. ft.	$....	$....	$ .11	$528.00
1¹⁄₄" Nails 4.8 M Pcs.	....	....	2.00	9.60
Corner Bead 96 L.F.	....	....	.07	6.72
Stop Bead 40 L.F.	....	....	.10	4.00
Joint Tape-8 rolls	....	....	.90	7.20

Item	Rate	Total	Rate	Total
Joint Compound - 6 5-gal cans	....	....	8.00	48.00
Cost for 4,800 sq. ft.		$....		$603.52
Cost per sq. ft		....		.13

Labor Placing Gypsum Wallboard.—When placing ½" gypsum wallboard in average size rooms, 2 carpenters experienced in wallboard erection should place about 3,000 sq. ft. of board per 8-hr. day, at the following labor cost per sq. ft.:

	Hours	Rate	Total	Rate	Total
Carpenter	16	$....	$....	$16.47	$263.52
Cost per sq. ft.			....		.09

Where gypsum wallboard is used to fireproof beams and columns, 2 carpenters should place about 1,200 sq. ft. per 8 hr. day at the following labor cost:

	Hours	Rate	Total	Rate	Total
Carpenter	16	$....	$....	$16.47	$263.52
Cost per sq. ft.			....		.22

Labor Finishing Gypsum Wallboard.—The finishing of gypsum wallboard is a multi-step process of tape coat, block coat, skim coat, and point-up coat. The necessity of sanding between the block and skim coats is not required if the finisher is careful and a first class mechanic.

The complete finishing labor for 1,000 sq. ft. of wallboard is as follows:

	Hours	Rate	Total	Rate	Total
Tape Coat	2.0	$....	$....	$16.47	$ 32.94
Block Coat	1.6	....	....	16.47	26.35
Skim Coat	1.3	....	....	16.47	21.41
Sand	0.4	....	....	16.47	6.59
Point-Up	0.8	....	....	16.47	13.18
Cost per 1,000 sq. ft.			$....		$100.47
Cost per sq. ft			....		.10

METAL STUDS AND FURRING

Metal Stud Partitions.—Runner channels are channel shaped members that are positioned at the top and bottom of the studs, as in top and bottom plates for a wood stud partition. The runner channels are attached to the floor and overhead structure with nails, screws, or powder actuated drive pins.

The metal studs, which are channel shaped with a backbend to give stiffness, are placed within the web of the runner channels, located for "on-center" spacing, plumbed, and screwed into place with a ⅜" pan head screw, top and bottom. The web of the metal studs have cutouts for the passage of conduit and piping.

Runner channels and studs are made in 1⅝", 2½", 3⅝", 4", 6" widths in 25 ga., 2½", 3⅝", 4", 6" widths in 20 ga., and 4", 6" widths in 18 ga. and 16 ga. metals.

The 25 ga. runners and studs are usually specified for non-load bearing partitions up to 14'-16' in height. Over this height and up to 22', the 20 ga. material is used. The 18 ga. and 16 ga. materials are used for greater heights, load bearing walls, and for exterior wall construction where lateral pressure (wind loads) will be imposed, therefore, requiring more stiffness and less deflection.

Metal Furring.—This material is made from 25 gauge metal in the following shapes:

1). Z furring channel is used in conjunction with rigid insulation board when furring exterior wall surfaces.

2). The hat shaped furring channel is used for furring walls and ceilings when insulation board is not required.

3). The resilient channel is used primarily on ceilings to provide a separation between the gypsum panels and the framing members for the floor above. It is used extensively in wood framed garden apartments for noise dampening between apartment units.

Estimating Material Quantities.

Example.—The following list of material quantities would be required for a metal stud partition 200 feet long, 10 feet high, stud at 24" on-center, 25 ga. metal, and installed with drive pins at 24" on-center, top and bottom to concrete slabs:

"Material Only"

Item	Rate	Total	Rate	Total
2½" 25 ga. Runner - 400 L.F.	$....	$....	$ .155	$ 62.00
2½" 25 ga. Stud - 1,010 L.F.	....	....	.17	171.70
1" Drive Pins - 202 ea.	....	....	.25	50.50
⅜" Pan Screws - 0.4 M.	....	....	4.50	1.80
Cost for 200 Lin. Ft		$....		$286.00
Cost per Lin. Ft.			....	1.43

Add extra studs for door and window openings and corners.

Labor to Install Metal Stud Partitions.—Using the above example given for materials, two (2) carpenters should erect the 200 lin. ft. of partition at the following labor cost:

	Hours	Rate	Total	Rate	Total
Layout Wall Line	1.0	$....	$....	$16.47	$ 16.47
Fasten Bottom Runner	4.0	....	....	16.47	65.88
Fasten Top Runner	4.0	....	....	16.47	65.88
Install Studs	3.0	....	....	16.47	49.41

	Hours	Rate	Total	Rate	Total
Plumb & Screw Studs	4.0	....	....	16.47	65.88
Cost for 200 lin. ft			$....		$263.52
Cost Per Lin. Ft			....		1.32

09300 TILE
CERAMIC WALL AND FLOOR TILE

Estimating Quantities.—Ceramic tile is estimated by the square foot, with trim pieces such as base, cap, etc. being estimated by the lineal foot.

Setting Methods.—Recent years have seen the development of many methods of adhering ceramic tile to a subsurface. There are, however, three methods which are most commonly used and accepted. These are :

Conventional Portland Cement Mortar Method. This method is to bond each ceramic tile with a layer of pure portland cement paste to a portland cement setting bed. This is done while the setting bed is still plastic. Wall tile must be soaked in water so that the water needed for curing is not absorbed from the paste. While being the traditionally accepted method, it is also more costly than the other methods.

Dry-Set Portland Cement Mortar Method. This method utilizes a dry curing portland cement mortar (accomplished through the use of water retaining additives) and has made ceramic tile installation cheaper and simpler. "Dry-set" is ideally suitable for use with concrete masonry, brick, poured concrete and portland cement plaster. It should not be used over wood or gypsum plaster. Labor costs are appreciably reduced when this method is used.

Water-Resistant Organic Adhesive. Organic adhesives can be used over smooth base materials, such as wallboards, plywood and metal. Labor productivity is comparable to that of 'dry-set' mortar.

Labor Productivity (sq. ft. per team day; team is 1 tile setter and 1 helper)			
Description	Face-mounted	Back-mounted	Unmounted
Glazed wall tile	....	75- 90 SF	55- 65 SF
Ceramic mosaic tile—walls	45- 50 SF	50- 65 SF	....
floors	100-125 SF	100-125 SF	....
Quarry tile floors	....	....	100-125 SF

FINISHES 209

Glazed wall tile	....	125-150 SF	100-125 SF
Ceramic mosaic tile—walls	100-125 SF	120-140 SF	....
floors	125-150 SF	125-150 SF	....
Quarry tile floors	....	....	125-150 SF
Glazed wall tile	....	150-175 SF	120-140 SF
Ceramic mosaic tile—walls	120-140 SF	125-150 SF	....
floors	150-175 SF	175-200 SF	....
Quarry tile floors	....	....	....
Cove Or Base	65-75 lin. ft.		
Cap	80-90 lin. ft.		

Placing Cement Floor Fill.—Cement floor fill under ceramic tile floors is usually placed by tile setters and helpers (one tile setter and one or two helpers working together). The fill is placed one or two days in advance of the tile if the overall thickness from rough floor to finished tile surface is over 3". For 3" thickness and under, fill and setting bed may be placed in one operation. When such tile floors are to be placed over wood subfloors it is necessary to first place a layer of water-proof building paper and a layer of wire mesh reinforcing before placing the fill.

A tile setter and two helpers should place 450 to 500 sq. ft. of fill per 8-hour day in areas large enough to permit efficient operations, at the following cost per 100 sq. ft.:

Material Costs

	Rate	Total	Rate	Total
6 bags portland cement	$....	$....	$ 4.00	$24.00
1 cu. yd. sand	....		12.00	12.00
Cost per 100 sq. ft.		$....		$36.00
Cost per sq. ft.		....		.36

Labor Costs

	Hours	Rate	Total	Rate	Total
Tile setter	1.5	$....	$....	$14.95	$22.43
Helper	3.0	....	....	12.54	37.62
Cost per 100 sq. ft.			$....		$60.05
Cost per sq. ft.			....		.60

Labor Setting 100 Sq. Ft. of Interior Marble

Description of Work	Lin. or Sq. Ft.	F'man Hours	Setter Hours	Helper Hours	Labor Hours	Hoist† Hours
Marble Toilet Stalls (per stall, not inc. backs)	....	2/5	3 3/10	3 3/10	2 3/5	1/8
Marble Toilet Stall Backs	Square	1 1/5	10	9	8	1/2
Marble Toilet Stall Partitions	Square	1	9	9	8	1/2
Marble Toilet Door Stiles	Lineal	2	18	18	8	1/2
Marble Cap on Stall Partitions	Lineal	1 1/2	12	12	8	1/2
Marble Door and Window Trim	Lineal	2	16	16	8	1/2
Marble Countertops	Square	1 1/5	12	10	8	1/2
Marble Thresholds, Each	Each	1/12	3/5	3/5	1/5	·· 1/4
Marble Plinths, 100 pcs.	Pcs.	4 1/2	36	36	8	1/2
Marble Balusters, 100 pcs.	Pcs.	5	40	40	16	1/2
Marble Balustrade, Handrail	Lineal	1 1/5	10	10	8	1
Marble Ashlar	Square	1 1/2	12	12	24	
Marble Column Bases to 3'x3'x1'	Each	1/10	1	1	2	1/16
Conc. Ft. Fill under Floors, 2 1/2" to 3"	Each	1/5	1 1/2	3		
Marble Floor Tile	Square		8	8	· 8	·· 1/2
Rub and Pol. Soft Marble Floors	Square	7/10	6**	6**		6*
Rub and Pol. Hard Marble Floors	Square	4/5	7**	7**		7*

*Elect. current for machine. **Machine operator. †Add hoist, engr. only if required.

Interior Marble

Labor Setting 100 Sq. Ft. of Interior Marble

Description of Work	Lin. or Sq. Ft.	F'man Hours	Setter Hours	Helper Hours	Labor* Hours	Hoist† Hours
Average labor cost for job, ⅞" stock		1 1/2	12	12	8	1/2
Average labor cost for job, 2" stock		1 1/2	16	16	12	1
Marble Base (short pcs.)	Lineal	1 1/2	12	12	8	1/2
Marble Base (long pcs.)	Lineal	1	9	9	8	1/2
Circular Base, double cost of straight work.						
3' Marble Wainscot or Die.	Square	1 1/2	11	11	8	1/2
3' to 4' Marble Wainscot or Die.	Square	1 1/5	10	10	8	1/2
4' to 5' Marble Wainscot or Die.	Square	1	9	9	8	1/2
5' to 7' Marble Wainscot or Die.	Square		8	8	8	1/2
Circular Wainscot or Die, double cost of straight work						
Wainscot Cap	Lineal	1 1/2	12	12	8	1/2
Marble Stair Treads, 3' to 3' 6"	Lineal	1 1/2	11	11	8	1/2
Marble Stair Treads, 4' to 6'	Lineal	1 1/10	9 1/2	9 1/2	8	1/2
Marble Stair Risers, 3' to 3' 6"	Lineal	1	11	11	8	1/2
Marble Stair Risers, 4' to 6'	Lineal	1 1/10	9 1/2	9 1/2	8	1/2
3'-6" Stair Wainscot on rake	Square	2 1/2	20	20	8	1/2
5' to 7' Stair Wainscot on rake	Square	1 1/2	14	14	8	1/2

*In the larger cities all marble is handled and distributed by helpers.
†Add hoisting engineer time only if required.

09400 TERRAZZO

Two methods are used in laying terrazzo floors over concrete construction. One is to bond it to the concrete and the other to separate it from the structural slab.

Labor Placing Terrazzo, Floor and Base.

The following are approximate quantities of various classes of terrazzo work a crew of 2 mechanics and 3 helpers should install per 8-hr. day:

Description of Work	No. Sq. ft. per 8-Hr. Day	Hrs. per 100 Sq. Ft. Mechanic	Helpers
Floor blocked off into 5'-0" squares	400–450	3.75	5.65
Floor blocked off into 4'-0" squares	375–425	4.00	6.00
Floor blocked off into 3'-0" squares	350–400	4.25	6.40
Floor blocked off into 2'-0" squares	325–375	4.60	6.90
Floor blocked off into 1'-0" squares	275–325	5.33	8.00
Floor border 12" to 24" wide	275–325	5.33	8.00

Terrazzo Cove Base	No. Lin. Ft.	Hours per 100 Lin. Ft.	
Terrazzo cove base, 3" high	60–65	13	13
Terrazzo cove base, 6" high	50–55	16	16

Double cove base, double time given for single base.

Rubbing and Finishing Terrazzo Work.— Terrazzo floors are rubbed by machine and by hand only in corners where the machine cannot reach.

On jobs consisting of large and small rooms, a man will rub and complete about 100 sq. ft. of terrazzo floor per 8-hr. day.

A man should rub and finish about 80 lin. ft. of terrazzo base per 8-hr. day.

09500 ACOUSTICAL TREATMENT

Acoustical tiles are made of many different materials, such as wood fiber, cane fiber, mineral wool, cork, specially processed mineral filaments, perforated metal units, and other insulating materials.

They are made in a variety of sizes, such as 12"x12", 12"x24", 24"x24", 24"x48", etc., and vary from ½" to 2" thick, depending upon the materials used.

Acoustical Tile Directly Applied.

The following quantities are based on 2 mechanics working together from a 4 to 6-ft. scaffold and laying tile in plain square or ashlar designs :

Size of Acoustical Tile	No. Pcs. Placed per 8-Hr. Day	No. Sq. Ft. Placed per 8-Hr. Day	Mechanic Hrs. per 100 Sq. Ft.
12"x12"	600-700	600-700	2.5
12"x24"	400-450	800-900	2.0

Acoustical Tile On Suspended Systems

Lay-in type systems for 2'x2' and 2'x4' grids will have a material cost of around 28¢ and 24¢ per sq. ft. respectively not including any channel supports if they are needed. One man should install 700 sq. ft. per eight hour day of 2'x4' grid, 600 sq. ft. for 2'x2' grid.

Concealed spline systems are more time consuming. Material costs will run around 30¢ a sq. ft. and one man will install 500 sq. ft. per day.

If channel carriers are required, the material cost will be about 8¢ per sq. ft. and one man should install about 500 sq. ft. per 8 hr. day.

Labor Cost To Install 100 Sq. Ft. 2'x2' Lay-in Grid

	Hours	Rate	Total	Rate	Total
Mechanic	1.3	$....	$....	$16.47	$21.41
Cost per sq. ft					.22

Labor Cost To Install 100 Sq. Ft. 2'x4' Lay-in Grid

	Hours	Rate	Total	Rate	Total
Mechanic	1.2	$....	$....	$16.47	$19.77
Cost per sq. ft					.20

Labor Cost To Install 100 Sq. Ft. 12 x 12 Z Bar System Grid

	Hours	Rate	Total	Rate	Total
Mechanic	1.6	$....	$....	$16.47	$26.35
Cost per sq. ft					.27

Labor Cost To Install 100 Sq. Ft. Channel Ceiling Supports

	Hours	Rate	Total	Rate	Total
Mechanic	1.6	$....	$....	$16.47	$26.35
Cost per sq. ft					.27

Acoustical Tile For Suspended Grid Systems.—This material is made "cut to size" ready to install into the grid system. 24"x24" and 48" boards will cost about 30¢ to 40¢ per sq. ft. depending on the type of tile, and one man should install about 100 sq. ft. per hour. The 12"x12" tiles for the Z bar system will cost about 50¢ to 65¢ per sq. ft. and one man should install about 100 sq. ft. in two (2) hours.

09550 WOOD FLOORING

Labor Laying 100 Sq. Ft. of Hardwood Floors

Description of Work	Laying Floors		
	Sq. Ft. Laid 8-Hr. Day	Carp. Hours	Labor Hours
Ordinary Workmanship			
25/32"x3 1/4" face softwood floors for porches, kitchens, factories, stores, etc.	400-500	1.8	0.6
25/32"x2 1/4" face Third Grade Maple, for warehouse, factory and loft building floors	375-425	2.0	0.6
25/32"x2 1/4" face Oak or Birch in residences, apartments, stores, offices, etc.	250-300	2.9	0.6
Same as above laid by experienced floor-layer	300-350	2.5	0.6
25/32"x1 1/2" face Third Grade Maple for warehouse, factory and loft building floors	300-325	2.3	0.7
25/32"x1 1/2" face Oak or Birch in residences, apartments, stores, offices, etc.	150-200	4.5	0.8
Same as above laid by experienced floor-layer	225-275	3.2	0.8
First Grade Workmanship			
25/32"x2 1/4" face Oak or Birch in fine residences, apartments, hotels, stores and offices	200-225	3.8	0.6
Same as above laid by experienced floor-layer	225-250	3.3	0.6
25/32"x1 1/2" face Oak or Birch, same class of work as described above	120-150	6.0	0.8
Same as above laid by experienced floor-layer	175-200	4.3	0.8

Labor Laying 1,000 Ft. B.M. Soft and Hardwood Flooring

Description of Work	Ordinary Workmanship		
	Ft. B. M. Laid Per 8-Hr. Day	Carp. Hours	Labor Hours
25/32"x3 1/4" face softwood floors for porches, kitchens, factories, stores, etc.	500-600	14.5	4
25/32"x2 1/4" face Third Grade Maple, used in warehouses, factory and loft buildings, etc.	500-575	15.0	4
25/32"x2 1/4" face Birch or Oak flooring in residences, apartments, stores, etc	350-400	21.0	4
Same as above laid by experienced floor-layer	400-450	19.0	4
25/32"x1 1/2" face Third Grade Maple, used in warehouses, factory and loft buildings, etc.	450-500	17.0	4
25/32"x1 1/2" face Birch or Oak flooring in residences, apartments, store and office buildings, etc.	250-280	30.0	5
Same as above laid by experienced floor-layer	350-400	21.0	5
First Grade Workmanship			
25/32"x1 1/4" face Oak or Birch flooring in fine residences, apartments, hotels, stores and offices	275-300	28	4
Same as above laid by experienced floor-layer	325-375	23	4
25/32"x1 1/2" face Oak or Birch, same class of work as described above for 21/4" face flooring	175-225	39	5
Same as above laid by experienced floor-layer	275-300	28	5

Estimating Quantities of Wood Flooring. — When estimating the quantity of board feet (b.f.) of wood flooring required for any job, take the actual number of sq. ft. in any room or space to be floored, and add allowances as given in the following tables :

Measured Size Inches	Finished Size Inches	No. of Pieces in Bundle	Add for Waste Percent	To Obtain Quantity of Flooring Repaired Multiply Area by	No. B.F. Flg. Required for 100 Sq. Ft. Floor
1x1	3/8 x 7/8	24	16 2/3	1 1/6 or 1.17	117
1x2	3/8 x1 1/2	24	33 1/3	1 1/3 or 1.33	133
1x2	3/8 x2	24	25	1 1/4 or 1.25	125
1x2 1/2	25/32x1 1/2	24	50 1/3	1 1/2 or 1.50	150
1x2 1/4	25/32x2	12	37 1/2	1 3/8 or 1.375	137 1/2
1x2 3/4	25/32x2	12	33 1/3	1 1/3 or 1.33	133
1x3	25/32x2 1/4	12	25	1 1/4 or 1.25	125
1x4	25/32x3 1/4	8			

Amount of Surface 1,000 Board Feet of Flooring Will Cover and Quantity of Nails Required to Lay it

How Measured	Size Flooring	Will Cover Sq. Ft. Fl.	Nailed Every	Lbs. Nails Req'd. Cut Nails	Helically threaded nails
1x2	3/8x1 1/2	750	8 in.	20 lbs. 4 "d" cut	60 lbs 7 "d" screw
1x2 1/2	3/8x2	800	8 in.	17 4 "d" cut	47 7 "d" screw
1x2 1/4	25/32x1 1/2	667	12 in.	70 8 "d" cut	
1x2 3/4	25/32x2	727	12 in.	56 8 "d" cut	54 7 "d" screw
1x3	25/32x2 1/4	750	10 in.	64 8 "d" cut	7 "d" screw
1x4	25/32x3 1/4	800	10 in.	29 8 "d" cut	24 7 "d" screw

Drill flooring for nails, if possible, for best results. No predrilling is required for helically threaded nails.

Note : The above figures are based on laying the flooring straight across a rectangular room without producing any design whatever.

Sanding Wood Floors by Machine

The class of workmanship of number of operations governs production and cost but where just an ordinary grade of workmanship is required in average size rooms in houses, apartments, offices, etc., a good machine and operator should finish and edge 800 to 900 sq. ft. per 8-hr. day, at the following labor cost per 100 sq. ft. :

	Hours	Rate	Total	Rate	Total
Operator-finisher ..	1	....	....	$16.47	$16.47
Cost per sq. ft			....		.17

Finishing Hardwood Floors

Labor Finishing Wood Floors.—Assuming the floor has been sanded and is ready for the seal, an experienced man can finish about 1,000 sq. ft. of natural finish floor per 8-hr. day. Colored seal takes more time as the excess pigment must be removed by hand wiping.

Labor Cost of 100 Sq. Ft. of Fast-Drying Seal and Lustre Finish Floor Finish on Wood Floors, Natural Color

	Hours	Rate	Total	Rate	Total
Mechanic..........	0.8	$....	$....	$16.47	$13.18
Cost per sq. ft			....		.13

PREFINISHED HARDWOOD PLANK FLOORING

Plank or strip flooring in ¾" thickness is available in several widths and finishes, most furnished in random lengths and widths. It is generally of red oak and predrilled for nailing. Widths are in the 3" to 8" range. Special distressed finishes and pegging are also available.

Plain flooring in narrower widths will run around $1.75 per sq. ft. For pegging add 10¢ per sq. ft. Wider planks will run $2.50 per sq. ft. Premium finishes can raise costs to as much as $2.75 a sq. ft.

One carpenter should lay at least 150 sq. ft. per day.

WOOD PARQUET FLOORING

Wood parquet is available in stock designs of oak, walnut, teak, cherry and maple. Standard thickness is 5/16", although heavier parquet is available on custom orders in 11/16" and 13/16" thicknesses. Sizes will vary with patterns running from 9"x9", 12"x12", 16"x16", on up to 36"x36".

Parquet is set in mastic on any level floor above grade. Most parquet is ordered factory finished, but may be unfinished if desired. Many patterns are available in "prime grade" and "character marked", square or bevelled edges.

Costs of all wood flooring are changing rapidly, but square foot material costs will approximate $1.70 for a standard 5/16" oak, $2.00 for walnut, and $2.70 for teak. The prime grades in more elaborate patterns will run much higher. If prefinished, add 30 cts per sq. ft.

Laying Wood Parquet Floors in Mastic Over Concrete.—Concrete subfloor shall be primed with one coat of asphalt priming paint and allowed to dry before laying parquet. Average covering capacity of primer 200 to 400 sq. ft. per gal. depending upon porosity of cement floor finish.

Labor Laying Wood Parquet Flooring.—When laying wood parquet floors in mastic in ordinary size rooms such as residences, apartments, etc., an experienced floorlayer with a helper should lay 300 to 350 sq. ft. of floor per 8 hr. day, while in large spaces such as schools, auditoriums, gymnasiums, etc., an experienced floorlayer should lay 350 to 400 sq. ft. of floor per 8-hr. day.

Laying Wood Parquet Floors in Mastic over Other Surfaces.—Wood parquet flooring may also be laid in mastic over plywood or board subflooring, present wood floors, asphalt tile, etc., using the same general procedures as given above for over concrete, except subsurface will not require priming before applying mastic.

Wood Block Industrial Flooring. Creosoted wood blocks 2" to 3" thick will run from $1.70 to $2.30 per sq. ft. One carpenter will install around 200 sq. ft. per day.

09650 RESILIENT FLOORING

Resilient flooring includes asphalt tile, rubber, cork, vinyl asbestos, pure vinyl tile and vinyl sheet flooring. Each type has its own qualities and should be selected on these considerations rather than cost alone. If material is to be installed on or below grade, cork tile and certain sheet vinyls are not suitable.

If grease is a problem, rubber, cork and asphalt tile perform poorly, while if resilience and quietness are important, cork and rubber tile are excellent, but sheet vinyl is poor.

A quick guide for approximate installed costs per sq. ft. is as follows :

Type	Thickness	Cost
Asphalt tile	3/32"-1/8"	$0.65-0.90
Vinyl asbestos	1/16"	0.80-0.90
	3/32"&1/8"	0.85-0.95
Cork tile-Prefinished	1/8", 3/16", & 5/16"	1.20-1.90
Vinyl tile	.050"	1.40-2.00
Vinyl sheet	.070"-.075	1.60-2.50
Rubber tile	1/8", 3/16"	1.50-1.85

Stair Treads

Size	Marbleized	Plain Black
12½"x36"x¼"	$ 9.75	$ 9.00
x42	11.50	10.50
x48	13.00	12.00

x54	14.50	13.50
x60	16.00	15.00
x72	19.50	18.00

Prices on Rubber Composition Top-Set Cove Base

Description	Price per Lin. Ft.
4" base..	$0.35
6" base..	.40

09900   PAINTING

The following are labor and material quantity tables for various types of painting operations for residential and commercial work.

Labor and Materials Required for Various Painting Operations

Exterior Work—Residential Description of Work	Painter No. Sq. Ft. per Hr.	Painter Hours 100 Sq. Ft.	Material Coverage Sq. Ft. per Gal.
Exterior Brush Painting Plain Siding And Trim			
Priming Coat	100-110	0.95	400-450
Second Coat	115-125	0.85	500-550
Third Coat	125-135	0.75	575-625
Exterior House Painting Rubberized Woodbond			
One Coat	125-135	0.75	450-500
Exterior Trim Only.			
Priming Coat	65-75	1.40	750-850
Second Coat	85-95	1.12	800-900
Third Coat	95-105	1.00	900-1,000
Oil Paint—Shingle Siding*			
First Coat	115-125	0.82	250-300
Second Coat	150-160	0.65	375-425
Oil Paint—Shingle Roofs*			
First Coat	100-120	0.90	120-150
Second Coat	155-175	0.60	220-250
Stain—Shingle Siding*			
First Coat	75-85	1.20	120-150
Second Coat	115-130	0.82	200-225
Stain—Shingle Roofs*			
First Coat	145-155	0.67	100-120
Second Coat	190-210	0.50	170-200
*If surfaces are exceptionally dry, increase labor hours and decrease covering capacity.			
Asbestos Wall Shingles			
First Coat	60-70	1.50	150-180
Second Coat	85-95	1.11	350-400
Brick Walls, Oil Paint			
First Coat	100-120	0.93	170-200
Second Coat	140-160	0.67	350-400
Third Coat	155-175	0.60	375-425
Porch Floors and Steps, Oil Paint			
First Coat	235-245	0.42	325-375
Second Coat	270-280	0.36	475-525
Third Coat	285-295	0.35	525-575

Labor and Materials Required for Various Painting Operations—Con't.

Exterior Work—Residential Description of Work		Painter No. Sq. Ft. per Hr.	Painter Hours 100 Sq. Ft.	Material Coverage Sq. Ft. per Gal.
Waterproof Cement Paint—Smooth Face Brick	First Coat	175-185	0.55	90-110
	Second Coat	260-270	0.38	140-160
Clear Waterproof Paint—Smooth Face Brick	First Coat	200-210	0.50	500-510
	Second Coat	200-210	0.50	590-610
Stucco, Medium Texture, Oil Paint	First Coat	90-100	1.10	340-360
	Second Coat	150-160	0.65	340-360
	Third Coat	150-160	0.65	340-360
Stucco, Medium Texture, Wpf. Cement Paint	First Coat	130-140	0.74	90-110
	Second Coat	190-210	0.50	125-145
Exterior Masonry, Stucco, Asbestos Cement Board and Shingle Siding—Rubberized Flat Finish	First Coat	100-120	0.91	325-375
	Second Coat	165-175	0.59	375-425
Concrete Walls, Smooth, Wpf. Cement Paint	First Coat	170-185	0.57	110-130
	Second Coat	275-285	0.36	150-170
Concrete Floors and Steps—Floor Enamel	First Coat	250-280	0.38	440-460
	Second Coat	190-210	0.50	575-625
	Third Coat	200-220	0.48	575-625
Cement Floors—Color Stain and Finish	First Coat	350-370	0.28	475-625
	Second Coat	280-300	0.35	450-500
Fences, Wire-Metal, Average	First Coat	100-110	0.95	900-1,000
	Second Coat	140-150	0.70	975-1,125
Shutters, Average, Each Coat	No. of Shutters	2-3	0.40	11-13
Downspouts and Gutters, Paint	First Coat	170-180	0.57	540-560
	Second Coat	185-195	0.53	575-600
Screens, Wood Only, Average, Each Coat	No. of Screens	6-8	0.14	45-55
Storm Sash, 2 Light, Average, Each Coat	No. of Sash	3-4	0.29	23-27

Spray Painting—Residential Work

Description of Work	Painter No. Sq. Ft. per Hr.	Painter Hours 100 Sq. Ft.	Material Coverage Sq. Ft. per Gal.
Brick, Tile and Cement—Latex Paint......First Coat	500-525	.20	90-110
Second Coat	525-550	.19	140-160
Brick, Tile and Cement—Oil Paint......First Coat	475-500	.21	250-275
Second Coat	500-525	.20	400-450
Rough Brick, Tile, Cement and Stucco*—Silicone......One Coat	800-900	.13	80-90
Asbestos Wall Shingles*......First Coat	300-325	.34	150-175
Second Coat	325-350	.31	225-250
Stucco*—Exterior Latex Paint......First Coat	400-425	.25	90-100
Second Coat	450-500	.22	125-150
Stucco*—Oil Paint......First Coat	300-325	.34	200-225
Second Coat	325-350	.30	350-400
Shingle Roofs—Oil Paint......First Coat	400-450	.25	125-150
Second Coat	450-500	.22	200-225
Shingle Roofs—Stain......First Coat	450-475	.22	125-150
Second Coat	475-500	.21	200-225
Shingle Siding*—Oil Paint......First Coat	350-375	.29	125-150
Second Coat	400-425	.25	225-250
Shingle Siding*—Stain......First Coat	400-450	.25	125-150
Second Coat	450-500	.22	200-225

*Trim Must be Figured Separately as Hand Work

Labor and Material Required for Interior Painting and Finishing

Finishing Interior Trim—Residential

Description of Work	Painter No. Sq. Ft. per Hr.	Painter Hours 100 Sq. Ft.	Material Coverage Sq. Ft. per Gal
Back Priming Interior Trim Up to 6" Wide ...... Lin. Ft.*	475-500*	0.21*	1,100-1,200*
Doors and Windows, Painting Interior ...... First Coat	140-150	0.70	475-500
Second Coat	115-125	0.85	500-550
Third Coat	115-125	0.85	500-550
Oiling or Priming Wood Sash ...... No. of Sash	9-10		
Preparatory Work for Enamel Finishes ...... Sanding	290-300	0.35	600-700
Sanding and Puttying	120-130	0.80	
Light Sanding	135-145	0.72	
Doors and Windows, Enamel Finish ...... Paint First Coat	140-150	0.70	575-600
Undercoat Second Coat	90-100	1.10	375-400
Enamel Third Coat	100-110	1.00	475-500
Add'l Third Coat	90-100	1.10	425-450
Four Coat Work. Add ½ Underc. ½ Enamel.			
Base, Chair Rail, Picture Mold, and Other Trim Up to 6" Wide. All Quantities are in Lineal Feet. ...... First Coat	250-260*	0.40*	800-900
Second Coat	145-155*	0.67*	1,100-1,200*
Third Coat	125-135*	0.77*	1,100-1,200*
Stain Interior Woodwork ...... One Coat	215-225	0.45	400-450
Stain and Fill Interior Woodwork, Wipe Off ...... One Coat	65-70	1.50	125
Shellac Interior Woodwork ...... One Coat	215-225	0.45	700-725
Varnish, Gloss, Interior Woodwork ...... One Coat	165-175	0.60	425-450
Varnish, Flat, Interior Woodwork ...... One Coat	170-180	0.57	600-625
Penetrating Stainwax—Standing Trim. ...... First Coat	170-180	0.57	525-550
Second Coat	190-200	0.51	600-625
Polishing Second Coat	190-200	0.51	

*Lineal Feet

Labor and Material Required For Interior Painting and Finishing—Cont.

Finishing Interior Trim—Residential

Description of Work		Painter No. Sq. Ft. per Hr.	Painter Hours 100 Sq. Ft.	Material Coverage Sq. Ft. per Gal.
Sanding for Extra Fine Varnish Finish	Sanding	40-45	2.25	
Synthetic Resin Finish, Requires Wiping	First Coat	190-200	0.50	550-600
	Second Coat	210-220	0.47	625-675
Spackling or Swedish Putty over Flat Trim	One Coat	60-65	1.60	140-150*
Glazing and Wiping over Enamel Trim	One Coat	60-65	1.60	1,050-1,100
Brush Stippling Interior Trim, Painted	Varnish	85-90	1.15	600-625
Flat Varnishing and Brush Stippling over Glazed Trim	Stipple	240-250	0.42	

Old Work

Description of Work		Painter No. Sq. Ft. per Hr.	Painter Hours 100 Sq. Ft.	Material Coverage Sq. Ft. per Gal.
Washing Average Enamel Finish	Washing	85-90	1.15	
Washing Better Grade Enamel Finish	Washing	60-65	1.60	
Polishing Better Grade Enamel Finish	Polish	170-180	0.57	1,700-1,800
Removing Varnish with Liquid Remover	Flat Surfaces	30-35	3.00	150-180
Wash, Touch Up, One Coat Varnish	Wash-Touch Up	165-175	0.60	
	Varnish	100-110	0.95	575-600
Wash, Touch Up, One Coat Enamel	Enamel	140-150	0.70	375-400
Wash, Touch Up, One Coat Undercoat and	Wash-Touch Up	70-75	1.40	
One Coat Enamel	Undercoat	140-150	0.70	475-500
	Enamel	75-85	1.25	475-500
Burning Off Interior Trim		25-30	4.00	
Burning Off Plain Surfaces		30-35	3.00	

*Per Lb.

Labor and Material Required For Painting Interior Walls

Description of Work	Painter No. Sq. Ft. per Hr.	Painter Hours 100 Sq. Ft.	Material Coverage Sq. Ft. per Gal.
Taping, Beading, Spotting Nail Heads and Sanding Gypsum Wallboard	90-110	1.00	600-700
Sizing, New Smooth Finish Walls .... Sizing	350-400	0.27	575-625
Wall Sealer or Primer on Smooth Walls .... Sealer	200-225	0.45	275-300
Wall Sealer or Primer on Sand Finish Walls .... Sealer	125-135	0.80	575-625
Smooth Finish Plaster—Flat Finish .... Sealer	200-225	0.45	575-625
Second Coat	165-175	0.60	500-550
Third Coat	175-185	0.55	575-625
Stippling	190-200	0.52	
Sand Finish Plaster, Flat Finish .... Sealer	125-135	0.80	405-500
Second Coat	135-145	0.70	400-450
Third Coat	125-135	0.80	425-475
Smooth Finish Plaster, Gloss or Semi-Gloss .... Sealer	200-225	0.45	575-625
Second Coat	170-180	0.57	500-550
Third Coat	140-150	0.70	500-550
Stippling	85-95	1.10	
Sand Finish Plaster, Gloss or Semi-Gloss .... Sealer	125-135	0.80	275-400
Second Coat	115-125	0.83	375-400
Third Coat	125-135	0.80	400-450
Texture Plaster, Average, Semi-Gloss .... Sealer	100-110	0.95	250-275
Second Coat	115-125	0.83	325-350
Smooth Finish Plaster—Latex Rubber Paints .... First Coat	125-135	0.80	375-400
Second Coat	140-160	0.67	300-350
Second Coat	140-160	0.67	300-350

Labor and Material Required For Painting Interior Walls—Con't.

Description of Work		Painter No. Sq. Ft. per Hr.	Painter Hours 100 Sq. Ft.	Material Coverage Sq. Ft. per Gal.
Sand Finish or Average Texture Plaster—Latex Rubber Paints	First Coat	110-130	0.83	250-300
	Second Coat	110-130	0.83	250-300
Glazing and Mottling over Smooth Finish Plaster		80-90	1.15	1,050-1,075
Glazing and Mottling over Sand Finish Plaster		60-65	1.60	875-900
Glazing and Highlighting Textured Plaster		80-90	1.15	825-850
Starch and Brush Stipple over Painted-Glazed Surface		115-125	0.83	
Flat Varnish and Brush Stipple over Painted-Glazed Surface		105-115	0.90	500-550
Texture Oil Paint over Smooth Finish Plaster	Size	215-225	0.45	675-725
	Texture	40-45	2.30	125-150
Water Texture over Smooth Finish Plaster	Size	215-225	0.45	700-725
	Texture	50-55	1.90	85-95
Latex Base Paint, New Smooth Plaster	First Coat	190-200	0.50	500-550
	Second Coat	210-220	0.47	650-700
On Rough Sand Finish Plaster	One Coat	150-160	0.65	325-350
On Cement Blocks	One Coat	135-145	0.70	300-325
On Acoustical Surfaces	One Coat	135-145	0.70	200-225

When using a roller applicator, increase quantities & reduce painter hours by 10-15%.

Labor and Material Required For Painting Interior Walls

Old Work — Description of Work	Painter No. Sq. Ft. per Hr.	Painter Hours 100 Sq. Ft.	Material Coverage Sq. Ft. per Gal.
Washing off Calcimine—Average Surfaces	115-125	0.83	
Washing Smooth Finish Plaster Walls—Average	145-155	0.67	
Washing Sand Finish Plaster Walls—Average	100-110	1.00	
Washing Starched Surfaces and Restarching, Smooth Surfaces ... Washing	145-155	0.67	
... Restarching	130-140	0.75	
Removing Old Wall Paper—Not Over 3 Layers	65-75	1.45	
Washing Off Glue After Removing Paper (including Fixing Average Cracks and Sizing)	125-135	0.77	
Cutting Hard Oil or Varnish Size Walls (including Fixing Average Cracks)	125-135	0.77	
Cutting Gloss Painted Walls (including Fixing Cracks)	125-135	0.77	
Washing, Touch Up, One Coat Gloss Paint to Smooth Plaster Surfaces ... Wash, Touch Up	125-135	0.77	
... One Coat	125-135	0.77	450-500
Synthetic-Resin Emulsion Paint over Old Painted Walls ... One Coat	190-200	0.50	400-425

Wallpaper, Canvas, Coated Fabrics, Paper Hanging, Wood Veneer

Description of Work	Unit	Painter No. Sq. Ft. per Hr.	Painter Hours 100 Sq. Ft.	Material Coverage Sq. Ft. per Gal.
Canvas Sheeting			1.82	
Coated Fabrics	Single Roll	50-60	2-2½	
Wallpaper—One Edge Work	Single Roll		3½-4	
Wire Edge Work	Single Roll		2½-3	
Butt Work	Single Roll		2¼-2¾	
Scenic Paper 40'x60'	Single Roll		1-1¼	
Wood Veneer Flexwood	Sq. Ft.	18		
Lacquer Finish Over Wood Veneer ... First Coat		100	1.00	275-325
... Second Coat		100	1.00	450-500

09950 WALLCOVERING

Wallpaper is estimated by the roll containing 36 sq. ft. Most rolls are 18" or 20½" wide and contain 36 sq. ft. or 4 sq. yds. of paper. The double roll contains just twice as much as the single roll, viz.: 72 sq. ft. or 8 sq. yds. of paper.

Where light or medium weight wallpaper is used, one gallon of paste should hang 12 single rolls of paper.

If heavy or rough texture paper is used, it will often be necessary to give it two or three applications of paste to obtain satisfactory results. On work of this kind one gallon of paste should hang 4 to 6 single rolls of paper.

There are prepared pastes on the market which require only the addition of cold water to make them ready for use.

One pound of prepared cold water dry paste should make 1½ to 2 gallons of ready to use paste.

Hanging Wallpaper on Walls and Ceilings, One Edge Work.—When hanging light or medium weight paper on ceilings and drops, a paper hanger should hang 28 to 32 single rolls of paper per 8-hr. day, at the following labor cost per roll :

	Hours	Rate	Total	Rate	Total
Paper hanger	0.27	$....	$....	$17.43	$4.71

Hanging Wallpaper on Walls, Butt Work.—Where light or medium weight paper is used for bed rooms, halls, etc., and where a good grade of workmanship is required with all paper trimmed on both edges and hung with butt joints, a paper hanger should trim, fit and hang 20 to 24 single rolls of paper per 8-hr. day, at the following labor cost per roll :

	Hours	Rate	Total	Rate	Total
Paper hanger	0.36	$....	$....	$17.43	$6.28

Vinyl Wall Covering—Vinyl wall covering is composed of a woven cotton fabric to which a compound of vinyl resin, pigment and plasticizer is electronically fused to one side. It comes 24", 27", and 54" wide in three weights : heavy (36 oz. per lineal yard); medium (24-33 oz. per lineal yard); and light (22-24 oz. per lineal yard).

A competent paper hanger can apply approximately 600 sq. ft. per 8 hr. day at the following costs per 100 sq. ft.

	Hours	Rate	Total	Rate	Total
Paper hanger	1.34	$....	$....	$17.43	$23.36
Cost per sq. ft.					.23

CHAPTER 10

MENSURATION

WEIGHTS AND MEASURES

Linear Measure

12	inches (in.) .... —1 foot .... ft.
3	feet (ft.) .... —1 yard .... yd.
16½	feet or 5½ yards .... —1 rod .... rd.
40	rods (rds.) .... —1 furlong .... fur.
8	furlongs (fur.) .... —1 mile .... mi.

Square Measure or Measures of Surface

144	square inches (sq. in.) ... —1 square foot .... sq. ft.
9	square feet (sq. ft.) .... —1 square yard .... sq. yd.
100	square feet (sq. ft.) .... —1 square .... sq.
30¼	square yards (sq. yds.) ... —1 square rod .... sq. rd.
160	square rods (sq. rds.) ..... —1 acre .... A.
640	acres (A.) .... —1 square mile .... sq. mi.

Cubic Measure

1,728	cubic inches (cu. in.) ... —1 cubic foot .... cu. ft.
27	cubic feet (cu. ft.) .... —1 cubic yard .... cu. yd.
128	cubic feet (cu. ft.) .... —1 cord .... cd.
24¾	cubic feet (cu. ft.) .... —1 perch (see Note) .... P.

A perch of stone is nominally 16½ ft. long, 1 ft. high and 1½ ft. thick, and contains 24¾ cu. ft. Many states west of the Mississippi River figure rubble by the perch containing 16½ cu. ft.

Avoirdupois Weight

437½	grains (gr.) .... —1 ounce .... oz.
16	ounces (ozs.) .... —1 pound .... lb.
100	pounds (lbs.) .... —1 hundredweight .... cwt.
2,000	pounds (20 cwt.) .... —1 ton .... T.

Dry Measure

2	pints (pt.) .... —1 quart .... qt.
8	quarts (qts.) .... —1 peck .... pk.
4	pecks (pk.) .... —1 bushel .... bu.

Liquid Measure

4	gills (gi.) .... —1 pint .... pt.
2	pints (pts.) .... —1 quart .... qt.
4	quarts (qts.) .... —1 gallon .... gal.
31½	gallons (gal.) .... —1 barrel .... bbl.
2	bbls. or 63 gals. .... —1 hogshead .... hhd.

SQUARE   RECTANGLE   PARALLELOGRAM

To Compute the Area of a Square, Rectangle or Parallelogram.—Multiply the length by the breadth or height. Example : Obtain the area of a wall 22 ft. (22'-0") long and 9 ft. (9'-0") high. 22x9=198 sq. ft.

To Compute the Area of a Triangle.—Multiply the base by ½ the altitude or perpendicular height. Example : Find the area of the end gable of a house 24 ft. (24'-0") wide and 12 ft. (12'-0") high from the base to high point of roof. 24 ft. x 6 ft. (½ the height)=144 sq. ft.

To Compute the Circumference of a Circle.—Multiply the diameter by 3.1416. The diameter multiplied by 31/7 is close enough for all practical purposes. Example : Find the circumference or distance around a circle, the diameter of which is 12 ft. (12'-0"). 12x3.1416=37.6992 ft. the distance around the circle or 12x31/7=37.714 ft.

CIRCUMFERENCE

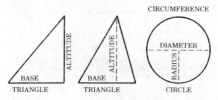

BASE   BASE   DIAMETER / RADIUS
TRIANGLE   TRIANGLE   CIRCLE

To Compute the Area of a Circle.—Multiply the square of the diameter by 0.7854 or multiply the square of the radius by 3.1416. Example : Find the area of a round concrete column 24 in. (24" or 2'-0") in diameter. The square of the diameter is 2x2=4. 4x0.7854=3.1416 sq. ft. the area of the circle.

The radius is ½ the diameter. If the diameter is 2 ft. (2'-0") the radius would be 1 ft. (1'-0"). To obtain the square of the radius, 1x1=1. Multiply the square of the radius, 1x3.1416=3.1416, the area of the circle.

To Compute the Cubical Contents of a Circular Column.—Multiply the area of the circle by the height. Example : Find the cubical contents of a round concrete column 2 ft. (2'-0") in diameter and 14 ft. (14'-0") long.

From the previous example, the area of a circle 2 ft. in diameter is 3.1416 sq. ft. 3.1416x14 ft. (the height)=43.9824 cu. ft. or for all practical purposes 44 cu. ft. of concrete in each column.

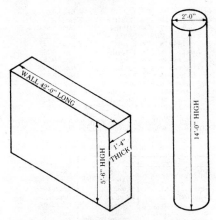

To Compute the Cubical Contents of any Solid.—Multiply the length by the breadth or height by the thickness. Computations of this kind are used extensively in estimating all classes of building work, such as excavating, concrete foundations, reinforced concrete, brick masonry, cut stone, granite, etc.

Example : Find the cubical contents of a wall 42 ft. (42'-0") long, 5 ft. 6 in. (5'-6") high, and 1 ft. 4 in. (1'-4") thick, 42'-0"x5'-6"x1'-4"=308 cu. ft. To reduce cu. ft. to cu. yds. divide 308 by 27, and the result is 1111/27 or 11½ or 11.41 cu. yds.

Table of Feet and Inches Reduced to Decimals

The following table illustrates how feet and inches may be expressed in four different ways, all meaning the same thing.

1 inch =	1 " =	1/12th	foot=0.083
1½ inches=	1½" =	1/8th	foot=0.125
2 inches=	2 " =	1/6th	foot=0.1667
2½ inches=	2½" =	5/24ths	foot=0.2087
3 inches=	3 " =	1/4th	foot=0.25
3½ inches=	3½" =	7/24ths	foot=0.2917
4 inches=	4 " =	1/3rd	foot=0.333
4½ inches=	4½" =	3/8ths	foot=0.375
5 inches=	5 " =	5/12ths	foot=0.417
5½ inches=	5½" =	11/24ths	foot=0.458

6 inches= 6	"= 1/2	foot	=0.5
61/2 inches= 61/2"	= 13/24ths	foot	=0.5417
7 inches= 7	"= 7/12ths	foot	=0.583
71/2 inches= 71/2"	= 5/8ths	foot	=0.625
8 inches= 8	"= 2/3rds	foot	=0.667
81/2 inches= 81/2"	= 17/24ths	foot	=0.708
9 inches= 9	"= 3/4ths	foot	=0.75
91/2 inches= 91/2"	= 19/24ths	foot	=0.792
10 inches=10	"= 5/6ths	foot	=0.833
101/2 inches=101/2"	= 7/8ths	foot	=0.875
11 inches=11	"= 11/12ths	foot	=0.917
111/2 inches=111/2"	= 23/24ths	foot	=0.958
12 inches=12	"= 1	foot	=1.0

Example : Write 5 feet, 7½ inches, in decimals. It would be written 5.625, which is equivalent to 55/8 feet.

Conversion Factors
S. I. Metric - English Systems

Multiply	by	to obtain
acres	0.404687	hectares
"	4.04687 X 10⁻³	square kilometers
acres	1076.39	square feet
board feet	144 sq. in. X 1 in.	cubic inches
" "	0.0833	cubic feet
bushels	0.3521	hectoliters
centimeters	3.28083 X 10⁻²	feet
"	0.3937	inches
cubic centimeters	3.53145 X 10⁻⁵	cubic feet
" "	6.102 X 10⁻²	cubic inches
cubic feet	2.8317 X 10⁴	cubic centimeters
" "	2.8317 X 10⁻²	cubic meters
" "	6.22916	gallons, Imperial
" "	0.2832	hectoliters
" "	28.3170	liters
" "	2.38095 X 10⁻²	tons, British shipping
" "	0.025	tons, U.S. shipping
cubic inches	16.38716	cubic centimeters
cubic meters	35.3145	cubic feet
" "	1.30794	cubic yards
" "	264.2	gallons, U.S.
cubic yards	0.764559	cubic meters
" "	7.6336	hectoliters
degrees, angular	0.0174533	radians
degrees, F (less 32 F)	0.5556	degrees, C
" C	1.8	degrees, F (less 32 F)
foot pounds	0.13826	kilogram meters
feet	30.4801	centimeters
"	0.304801	meters
"	304.801	millimeters
"	1.64468 X 10⁻⁴	miles, nautical
gallons, Imperial	0.160538	cubic feet

"	"	1.20091	gallons, U.S.
		4.54596	liters
gallons, U.S.		0.832702	gallons, Imperial
"	"	0.13368	cubic feet
"	"	231.	cubic inches
"	"	0.0378	hectoliters
"	"	3.78543	liters
grams, metric		2.20462×10^{-3}	pounds, avoirdupois
hectares		2.47104	acres
		1.076387×10^{5}	square feet
hectares		3.86101×10^{-3}	square miles
hectoliters		3.531	cubic feet
"		2.84	bushels
"		0.131	cubic yards
hectoliters		26.42	gallons
horsepower, metric		0.98632	horsepower, U.S.
horsepower, U.S.		1.01387	horsepower, metric
inches		2.54001	centimeters
"		2.54001×10^{-2}	meters
"		25.4001	millimeters
kilograms		2.20462	pounds
		9.84206×10^{-4}	long tons
		1.10231×10^{-3}	short tons
kilogram meters		7.233	foot pounds
kilograms per m		0.671972	pounds per ft
kilograms per sq cm..		14.2234	pounds per sq in
kilograms per sq m..		0.204817	pounds per sq ft
		9.14362×10^{-5}	long tons per sq ft
kilograms per sq mm		1422.34	pounds per sq in
" " "		0.634973	long tons per sq in
kilograms per cu m....		6.24283×10^{-2}	pounds per cu ft
kilometers		0.62137	miles, statute
"		0.53959	miles, nautical
"		3280.7	feet
liters		0.219975	gallons, Imperial
"		0.26417	gallons, U.S.
"		3.53145×10^{-2}	cubic feet
"		61.022	cubic inches
meters		3.28083	feet
"		39.37	inches
"		1.09361	yards
miles, statute		1.60935	kilometers
" "		0.8684	miles, nautical
miles, nautical		6080.204	feet
" "		1.85325	kilometers
" "		1.1516	miles, statute
millimeters		3.28083×10^{-3}	feet
"		3.937×10^{-2}	inches
pounds, avoirdupois ..		453.592	grams, metric
" "		0.453592	kilograms
" "		4.464×10^{-4}	tons, long
" "		4.53592×10^{-4}	tons, metric

Conversion Factors (Cont'd)

Multiply	by	to obtain
pounds per ft	1.48816	kilograms per m
pounds per sq ft	4.88241	kilograms per sq m
pounds per sq in	7.031×10^{-2}	kilograms per sq cm
" " " "	7.031×10^{-4}	kilograms per sq mm
pounds per cu ft	16.0184	kilograms per cu m
radians	57.29578	degrees, angular
square centimeters	0.1550	square inches
square feet	9.29034×10^{-4}	acres
square feet	9.29034×10^{-6}	hectares
" "	0.0929034	square meters
square inches	6.45163	square centimeters
" "	645.163	square millimeters
square kilometers	247.104	acres
" "	0.3861	square miles
square meters	10.7639	square feet
" "	1.19599	square yards
square miles	259.0	hectares
" "	2.590	square kilometers
square millimeters	1.550×10^{-3}	square inches
square yards	0.83613	square meters
tons, long	1016.05	kilograms
" "	2240.	pounds
" "	1.01605	tons, metric
" "	1.120	tons, short
tons, long, per sq ft	1.09366×10^{-4}	kilograms per sq m
tons, long, per sq in	1.57494	kilograms per sq mm
tons, metric	2204.62	pounds
" "	0.98421	tons, long
" "	1.10231	tons, short
tons, short	907.185	kilograms
" "	0.892857	tons, long
" "	0.907185	tons, metric
tons, British shipping	42.00	cubic feet
" " "	0.952381	tons, U.S. shipping
tons, U.S. shipping	40.00	cubic feet
" " "	1.050	tons, British shipping
yards	0.914402	meters